£5·40

3

D0305687

An introduction to engineering economics

19184.

An introduction to engineering economics

The Institution of Civil Engineers

Printed by William Clowes and Sons Limited, London and Beccles

Preface

The postwar years have witnessed an increasing interest among engineers in the subject of economics. This interest has been accompanied by a growing awareness of the important part which the subject plays in their everyday affairs and of the need to make the best use of our limited national resources.

Although the present century has seen almost every aspect of economics covered by a steadily increasing volume of literature there has hitherto been a scarcity of work written specifically with the intention of providing a bridge between economic theory and engineering practice.

To remedy this situation the present handbook has been produced to introduce the engineer to accepted economic principles of evaluation and to demonstrate the application of current techniques to the solution of problems of the kind which the engineer is likely to encounter during the course of his professional work.

Today the preparation of an economic evaluation of a project as a basis for investment decision is of increasing importance. The basic principles of such evaluations are outlined in the following chapters but those who require a more detailed exposition of the subject should refer to some of the works listed in the bibliography.

COMMITTEE FOR THE REVISION OF
An introduction to engineering economics

Chairman: R. le G. Hetherington, OBE, MA, FICE
K. G. Armstrong, BE, FICE
J. W. Baxter, BSc(Eng), FICE
D. J. Bolton, MSc, FIEE
F. E. Bonner, BSc(Econ), DPA, FCA, JDipMA, FIMTA
N. Borg, FICE
Professor H. C. Edey, BCom, FCA
J. K. Hunter, BSc(Eng), FICE
J. E. G. Palmer, MA, FICE
R. E. J. Worth, BSc(Eng), FICE
Secretary: P. B. E. Thompson, BSc(Eng), FICE

DRAFTING SUB-COMMITTEE

Chairman: J. K. Hunter, BSc(Eng), FICE
K. G. Armstrong, BE, FICE
B. V. Carsberg, MSc(Econ), ACA
Professor H. C. Edey, BCom, FCA

The Council of the Institution is greatly indebted to Professor Edey and Mr
Carsberg of the London School of Economics for their assistance in the preparation of this handbook.

Production editor: T. L. Dennis

Contents

viii

APPENDIX F: COMPOUND INTEREST AND ANNUITY TABLES 169

Chapter 1

INTRODUCTION

Engineers form one of the principal spending professions in the sense that they carry much of the responsibility for the wise use of those sources of wealth which provide the material basis of an expanding civilization. Although much of the engineer's work is concerned with the design of projects which form part of a policy which has already been decided, the principles set out in this handbook are nevertheless relevant to all his work and are especially important when major policy decisions are involved.

Apart from the land itself, the most important natural resources include climate and water, the fertility of the soil and the presence of useful minerals. To these may be added a topography which favours easy communications. The development of these natural resources is dependent upon the application of human skill and labour supported by man-made capital resources, themselves the result of previous productive effort. These resources represent the potential for a high standard of living and at any moment their availability is limited in the sense that if a resource is used for one purpose it is denied to another which must therefore be forgone. Thus, competition arises between desirable projects.

Economic analysis as applied to engineering is concerned with assessing the real cost of using resources in order to establish priorities between competing proposals. Its purpose is to provide the engineer with a means of judging the relative economic merits of alternative schemes and of ensuring that available resources shall be used to achieve the desired end with the minimum expenditure of means.

For the most part the decision whether to employ resources for one purpose rather than for another lies not with trained engineers but with politicians, administrators, financiers, bankers and others, many of whom

1

are familiar with the techniques of economic appraisal. If they are to make a wise choice between competing programmes of expenditure they must be properly informed by those who can not only assess the comparative technical merits of alternative courses of action, but can also evaluate their relative real costs when expressed in terms of the benefits to be secured.

It has been claimed (World Bank, Policies and Operations, June 1960) that the 'success of modern business is as fully dependent on financial planning and control, marketing, sound organizational structure, adequate supervision and training of personnel as it is upon machines and technology'.

In thus emphasizing the importance of some of the wider aspects of economic activity, the attention of engineers is drawn to the need for breadth of vision and a better understanding of all the factors which contribute to a successful enterprise. Thus it may be said that the main purpose of this handbook is to broaden the horizon of the engineer and to equip him to make a greater contribution to the conduct of affairs.

Since much of an engineer's work is concerned in making a reasoned choice between possible alternatives, he must be able to apply an appraisal technique which will lead to an evaluation of their real costs. A vital element in such an evaluation is the time pattern of the expenditure of resources in achieving a fixed objective and of the flow of benefits arising therefrom. Closely associated with time is change, one of the dominant characteristics of the modern world, which is continually affecting our economic life. Perhaps the most important source of change is technological advance, the direction of which may sometimes alter unexpectedly, creating new demands and often causing a decline in traditional activities.

ECONOMIC EVALUATION AND FINANCIAL PLANNING

It is useful to differentiate between economic evaluation and financial planning. Economic evaluation employs accepted principles and procedures to determine the real cost to the community or to an organization, usually within a defined range of possibilities, of using resources to achieve a specific purpose. It is less concerned with the earnings and revenues necessary to meet the obligations of a loan than with those which are

2

necessary to justify the selection of a particular project in the prevailing economic environment. Such an evaluation must take into account the need for carrying out related investment projects and where necessary must examine the impact which the project is likely to have on the regional or national economy. Studies might include, for example, the effect of using foreign exchange for the importation of goods and services and of import duties or subsidies which may have a significant and distorting effect on costs. In some cases too it may be necessary to examine the influence which the project is likely to have on employment, either locally or nationally.

Part II is concerned with the general principles of project appraisal while Part III is devoted to the application of project appraisal techniques which are illustrated by a number of numerical examples typical of the kind of problem which falls within the ambit of the engineering profession.

Financial planning is concerned with the flow of money arising from an accepted project and involves the kind of appraisal which a bank would make in assessing the prospects of an undertaking being able to generate sufficient revenues to repay a loan on the due date. Its objective is specific, not comparative, and it employs firm rules in order to demonstrate the financial results of judgements made earlier in the economic evaluation.

Part IV deals with budgetary problems and financial planning including provision for the repayment of loans. Also dealt with in Part IV is the allocation of costs in multi-purpose projects.

NEED FOR JUDGEMENT

Planning is concerned with the future, but to plan at all it is necessary to make assumptions as to what the needs of the future are likely to be. Complex procedures are now used in the preparation of such forecasts; some of these involve other disciplines than engineering and may include the use of simulation models and the like. But however complete our records of the past may be, and however refined are our methods of using them, we should recognize that no amount of data will remove the fundamental uncertainties which surround any attempt to peer into the future. Thus it follows that any conclusion reached on the evidence should be looked upon as representing no more than the best conclusion which can be arrived at on the information available; a conclusion which is increasingly likely to prove wrong with the passage of time. However complete the information, the need for personal judgement and experience in making economic decisions remains. Moreover, it must be recognized that both the selection of evidence and the choice of assumptions are to some extent subjective as are the conclusions drawn from them. It follows therefore that the application of refined methods of evaluation will not ensure reliable results if the basic data used are unsound or the assumptions poorly selected. It must be accepted that future changes will probably make it clear that the decisions taken were not, in fact, as good as they once appeared to be, and that with greater foresight a different course might have been pursued with better results. Notwithstanding these limitations there is no escape from the need to employ careful economic analysis based on projections of the past.

Closely allied to technological change is the discovery not only of new resources but also of new uses for old resources. The existence of coal and mineral oil deposits has been known since early historical times but it was not until the latter part of the eighteenth century that they began to be put to substantial economic use. Although the element uranium was first isolated in 1842 its existence was unheeded by the general public for over a hundred years. It then assumed dramatic importance and its potential as a future source of energy began to be understood.

OVERSEAS ENGINEERING WORK

During the latter part of the nineteenth and the early part of the twentieth century British engineers played a predominant role in assisting private enterprise, often supported by the governments of the territories concerned, to open up and develop many parts of the world. This work, much of which was concentrated on the provision of communications and transport: harbours, railways and roads, provided the essential foundations for economic growth and stimulated industrialization. However, by the end of the First World War many of the countries which had formerly welcomed British engineering talent had become in some measure self-supporting in respect of technical expertise: a trend which was greatly accelerated during the inter-war years.

More recently the drying-up of these opportunities has been replaced by a new flow of financial and technical assistance to the emergent nations, and this has stimulated a demand for the provision of engineering services. Since the Second World War the demand for such services has steadily grown and by 1966 two-thirds of the value of the overseas engineering work for which British engineers were responsible lay in such regions.

The problems to be faced in such countries tend to be different both in kind and in scale from those encountered in more highly developed countries. Some of the more important aspects of development planning are dealt with in Appendix C

USE OF SCARCE RESOURCES

If we consider the world as a whole, it is clear that the resources available are limited and are insufficient to meet all the wants of mankind. However, to a group of individuals or a company a particular resource is not necessarily restricted in an absolute sense, since their requirements may well represent but a small part of the total world supply and, by offering a higher price, they may be able to bid this resource away from others.

In the short run, however, even such a group may find that the supply of certain resources is absolutely limited. For example, the availability of labour may be restricted locally and it may be impracticable to import labour from other areas.

Scarcity of resources thus shows itself in two ways. In the first place money is necessary to acquire the resource and if the benefits to be derived from its employment do not accrue at the same time as the expenditure is incurred, then the money, if not already available, will have to be raised. Hence the availability of money may impose an effective limit on ability to acquire resources.

In the second place, and more fundamentally, the supply of particular resources is restricted either as a result of natural limitation as in the case of land, or as in the case of manufactured goods because productive capacity at any given time is limited.

It is sometimes convenient to classify projects either as capital or non-capital projects. Those falling into the former category are distinguished by the longer time which elapses between the outlay and the receipt of

5

benefits. Where a lapse of time occurs, the owner of the capital employed, whether an individual or the community at large, will require to be compensated for waiting. This compensation is expressed in terms of the interest cost which must be associated with such projects, and a capital project may therefore be defined as one where the element of interest cost is significant.

The aim of economic evaluation is to secure the greatest benefit from the resources available. The choice may often require consideration of alternative ways of carrying out a particular project, such as, for instance, achieving an optimum balance between the use of labour and machines. It will be clear that where labour is scarce, and the cost is therefore high, it will be more economic to employ more machines until the cost of increasing the use of machines equals the cost of employing more persons.

The methods of economic evaluation described in succeeding chapters are of general application and can be used with equal validity whether, say, the problem is to determine the scale on which a manufacturing activity should be carried on, or to establish the justification for undertaking a large regional development project. The main differences which arise in appraising inherently different kinds of projects are those concerned with the measurement of the flow of costs and benefits over time. Some of these costs and benefits can readily be assessed in terms of money. In other cases it is difficult to attach money values, as with projects undertaken by the state or public authorities where the benefits may be in the form of a service which is not sold but is provided free.

The social benefits arising from expenditure in the public sector are often an important element in the justification of such projects. It must therefore be emphasized that the absence from this handbook of any explanation of the techniques which may be used for estimating such benefits should not be taken as evidence that this aspect of cost–benefit analysis can safely be ignored. However, in many cases no really satisfactory method of making such estimates has yet been found and the final decision rests on political judgement.

Finally, it cannot always be assumed, even when due weight has been given to the more obvious social benefits attached to a project, that the choice will not be affected by even broader considerations such as national security.

Chapter 2   Discounted cash flow methods

THE NATURE OF PRESENT VALUE CALCULATIONS

Most people given a choice between receiving a specified sum (say, £100) now and the same amount at a later time (say, after one year) will express a preference for the present sum.   One reason for this preference is that money (cash) received now can be invested to earn interest (or, what comes to the same thing, used to repay a loan and thereby save interest); it will then amount to a larger sum in one year's time.   For example, the sum of £100 invested at 5% per annum will amount to £105 after one year.

Suppose that a choice has to be made between the receipt of £100 now and £102 after one year.   If the investment of £100 for one year could be made at a rate of 5% p.a. it would yield £105; thus, the acceptance of £100 now may be said to yield a net surplus of £3 (£105 − £102) in one year's time. Its acceptance is better by £3 than the alternative of accepting the later receipt.

The same result can be obtained in a slightly different way.   Suppose it is asked what sum would have to be set aside now, in order to yield £102 after one year, given the rate of interest of 5% p.a.   The answer is £102/1·05, that is, £97·1, and this amount is known as the present value or present worth of £102 in one year's time.   A comparison between £97·1 and £100 (the latter is already a present value because it arises immediately) again indicates that the choice of £100 now is to be preferred; £100 is larger than the amount which would have to be set aside now to yield £102 after one year.

Thus it can be seen that a choice between benefits which arise at different times should not be made by comparing their absolute amounts.   Instead, the actual receipts should be converted into their equivalent values as measured at a single point of time (and it is usually convenient to use the

2

present time, that is, the time at or near to commencement of the plan under review) in order to derive amounts which may be compared directly. The same considerations apply in comparing costs which arise at different times.

Suppose the relevant annual rate of interest, expressed as a decimal, is r, and that it is to be compounded with annual rests, that is to say, one year's interest is added at the end of each year. Then the sum of £a_0 arising at time 0 is equivalent to £$a_0(1+r)^n$ in n years' time. Similarly, the sum of £a_n, arising in n years' time, has a present value of £$a_n/(1+r)^n$.

The process of calculating a present value in this way is known as discounting and the terms interest rate and discount rate tend to be used interchangeably.

NET PRESENT VALUE OF A PROJECT

The application of this principle to the appraisal of a capital project is illustrated in Example 2.1.

EXAMPLE 2.1

It is assumed that a project will be acceptable if it leads to an increase in the present value of the resources of the undertaking (or the individual) making the decision. Acceptance of the project involves making an immediate cash outlay of £1000 and receiving:

£400 at the end of year 1
£500 at the end of year 2
£300 at the end of year 3

The rate of interest is 6% p.a. compounded annually.
Present values taken at the commencement of the project are as follows.
Present values of receipts:

after 1 year	£400/1·06	£377·4
after 2 years	£500/1·06²	£445·0
after 3 years	£300/1·06³	£251·9
		£1074·3
less present value of outlay		£1000·0
net present value		£74·3

The net present value (that is, the present value of receipts less the present value of outlays) is positive. This indicates that the resources of the undertaking will increase in value if the project is accepted. This may be seen more clearly if the example is extended slightly.

Suppose that the cash required for the initial outlay and for any further amounts required for other purposes will be borrowed on bank overdraft and that all receipts will be paid over to the bank when they arise, in reduction of the overdraft. Let the rate of interest charged on the overdraft be 6% p.a. compounded annually. The present value calculation indicates that the undertaking can borrow an additional £74·3 at the start of the project, and use it in any way it likes, because the receipts from the project will be sufficient to repay the total amount borrowed with interest. Thus, if the project is accepted, the undertaking can increase the money value it has available for immediate disposal by £74·3. The overdraft account would run as follows.

Start of year 1	borrow initial outlay	£1000·0
	borrow additional sum	74·3
		1074·3
End of year 1	add interest—6% on £1074·3	64·5
		1138·8
	less cash receipt	400·0
		738·8
End of year 2	add interest—6% on £738·8	44·3
		783·1
	less cash receipt	500·0
		283·1
End of year 3	add interest—6% on £283·1	16·9
		300·0
	less cash receipt	300·0
		0

Methods which involve the discounting of estimated cash flows as in the above example, or the use of compound interest formulae in ways that are mathematically equivalent, are often referred to as discounted cash flow (DCF) methods. Various other methods of investment appraisal are sometimes used. However, those which are not mathematically equivalent to a net present value method can be shown to suffer from shortcomings under most conditions. Some of the more common methods are discussed in Appendix A.

EFFECT OF UNCERTAINTY

It is convenient at this point to make a brief reference to the effect of uncertainty on project appraisal, a subject which is discussed in greater detail in Chapter 6. However carefully the estimates are made the actual cash flows may be different. In practice, therefore, some estimates will usually be made of possible alternative outcomes. If all the alternative outcomes which are thought reasonably possible show positive present values the project may be accepted on the ground that it is certain to be worthwhile. However, if some of these alternatives show negative present values (i.e., the present value of payments exceeds that of receipts) it may be reasonable to reject the project. Whether it is rejected or not will depend partly upon the estimated chance that an unfavourable outcome will arise and upon the size of the latter. It will also depend on the attitude of the undertaking to different estimated surpluses and deficits. For instance, a deficit of a given magnitude might be thought disastrous (it might for example put the undertaking into bankruptcy) while a surplus of the same size might be viewed as only moderately beneficial. Thus the acceptance or rejection of a project in conditions of uncertainty will depend, in part, on the relative weight which the undertaking attaches to the possible surpluses and deficits above or below the expected outcome. In the examples in this chapter it is however assumed for simplicity that the outcome of a project can be assessed with sufficient certainty for the possibility of alternative results to be ignored.

FINANCIAL OBJECTIVES

The significance of the net present value calculation in the above example arose from the assumption that the project was financed by a bank overdraft. However, projects are not usually financed wholly or even mainly in this way. In a more realistic situation the significance of a net present value calculation requires a more extended explanation.

10

It is convenient to start by defining the financial objectives of the undertaking since the purpose of the appraisal techniques employed is to reveal whether acceptance of a project is likely to lead to results which are consistent with these objectives. Although the discussion will be based on an undertaking in the private sector of the economy the same principles can be applied to the public sector. In the latter case special considerations may arise as explained in Chapter 4.

In order to carry on its operations an undertaking uses resources provided by various people. It hires labour and managerial skill in return for the payment of wages and salaries. It requires finance for this purpose and also to enable it to buy or hire other resources such as land, buildings, machinery, materials. Finance includes not only direct but indirect money provision as when a supplier allows payment due to him to remain outstanding or an 'owner' refrains from withdrawing profit from an enterprise. Finance is required because of the need to bridge the interval between the time when resources are acquired and the time when they yield money benefits. Finance may be supplied by short-term or long-term creditors and by owners such as the ordinary shareholders of a company. The sources of finance are described in more detail in Appendix B.

Investment appraisal in the private sector is normally based on the idea that a company is owned by its ordinary shareholders who provide finance in the hope of gain but without any certainty of what their return may be. The surplus available after meeting the costs of running the undertaking, including payment of interest on borrowed money, and after payment of taxation and dividends to preference shareholders, is held by the directors of the company for the benefit of the ordinary shareholders. Since the ordinary shareholders are, in effect, the legal owners of the company it has been accepted as appropriate that the financial aim of the company should be to maximize the value of their interest in the company, that is, maximize the present value of the future net benefits expected to accrue to them from the company. Similarly, the financial aim of an organization in the public sector may be said to be to maximize the value of the net benefits it is expected to provide for the community. In both cases these objectives are subject to normal social constraints, that is, the maximization is expected to be within a framework of acceptable social behaviour.

SIGNIFICANCE OF PRESENT VALUE CALCULATIONS

Suppose an organization is considering a number of possible investments and can obtain, at a price, enough resources in the form of labour, finance, and so on, to enable it to undertake as many of them as it wishes. Suppose also that all the required finance will be provided by ordinary shareholders. Consider a particular investment project for which a statement has been prepared giving details of the estimated initial cash outlay and subsequent cash receipts.

EXAMPLE 2.2

It is estimated that the project will require an initial cash outlay of £10 000. It will earn cash receipts of £4000, £5000 and £4000 at the end of the first, second and third years and then come to an end. The £10 000 required for the initial outlay will be paid into the enterprise by the ordinary shareholders and the subsequent net cash receipts will be used to make dividend payments to them. (The conditions of dividend payment imposed by law are assumed to be satisfied.)

Cash receipts exceed the outlay and the shareholders will receive £3000 more than they contribute. This might suggest the project should be accepted. However, while an organization should normally reject a project which requires a cash outlay in excess of the sum of the expected cash receipts, it should also reject projects which yield a cash surplus if the investor could put his money to some better alternative use. The justification for using an investor's money for a particular capital project is the expectation that such use will enable him to receive a stream of benefits at least as attractive to him as any he could hope to obtain by employing his money in some other way. One alternative open to him may be an investment in another undertaking. Another is the opportunity to spend the money on increased present consumption. If he uses cash for any investment he thereby deprives himself of the opportunity for immediate consumption. It must be supposed that as an inducement to invest he must be offered the prospect of increasing his capital resources by an amount that will at least compensate for whatever lost opportunity he rates most highly.

It is convenient to standardize the measure of such an increase by using a rate of interest compounded annually. Suppose the rate of interest required to justify investment in the project is 10% p.a. The present value of the project may be calculated, in £s, as

$$-10\ 000+4000/1{\cdot}10+5000/1{\cdot}10^2+4000/1{\cdot}10^3$$
$$= -10\ 000+3636+4132+3005$$
$$= 773$$

The fact that the net present value is positive indicates that throughout its life the project earns on the amount of the original investment not yet recovered (an amount which falls from year to year) a return in excess of the required 10% p.a., thus satisfying the test of acceptability. This can be demonstrated arithmetically by a calculation similar to that used previously in the bank overdraft illustration (Example 2.1).

The estimation of the appropriate rate of interest to be used is a matter of considerable practical difficulty. It is considered later in this chapter and in Appendix B. But once the rate has been established the procedure described may be applied in turn to each project under consideration in order to determine its acceptability.

INTERDEPENDENT PROJECTS

In the preceding discussion it was assumed that each project could be appraised independently without considering what other projects were available and whether they were worthwhile.

In some situations, however, the worthwhileness of one project can only be assessed by comparing its expected results with those of other projects. There are two reasons for this.

Mutually exclusive projects
In the first place, some projects are mutually exclusive in the sense that acceptance of one alters the willingness of an undertaking to accept another. For example, if a manufacturer requires a new warehouse he may consider a number of possible alternatives: leasing compared with construction, different sites, etc. If he only requires one warehouse the set of alternatives open to him constitutes a series of mutually exclusive projects.

Increasing cost of finance
In the second place, progressive increases of capital may be associated with higher costs, that is to say, the return required on behalf of those

13

providing finance may increase with the amount of money which is raised in a given period. In these circumstances the acceptance of one project makes the acceptance of another less attractive because a higher rate of return on the additional capital is expected.

An extreme example of this kind of interdependence arises when only a limited number of projects can be accepted because the amount of capital which can be raised cannot, at any cost, be increased beyond a certain limit. When this is the case an investment project has to satisfy a double test. It must promise earnings which will give as much satisfaction to the providers of finance as they could obtain from any use of the same funds outside the organization. It must also promise to yield returns which have as great a value as those of any internal investment which would have to be forgone if the project under consideration were accepted.

APPRAISAL OF INTERDEPENDENT PROJECTS

Consider again Example 2.2. Had the required initial outlay been £10 773 instead of £10 000 the project would have had a zero net present value and it would have been judged to be only just worthwhile; the organization would have been at the point of indifference between acceptance and rejection. Thus, given the 10% p.a. rate of interest, the investors may be said to derive the same satisfaction from having £10 773 available now as they would derive from the prospect of cash receipts of £4000, £5000 and £3000 after 1, 2 and 3 years, and since the project actually required an outlay of only £10 000, its acceptance may be said to add £10 773 − £10 000 = £773 to the value of the capital resources of the investors. Thus the net present value of a project may be used as an estimate of the value which would accrue to the ordinary shareholders of a company as the result of accepting the project. It follows that the best of a set of mutually exclusive projects is the one with the highest net present value. Example 2.3 illustrates how projects should be selected in the type of situation in which the cost of capital increases with the amount which has to be raised.

EXAMPLE 2.3

Suppose an organization has a number of investment projects available each requiring an outlay of £100 and each lasting exactly one year. It can raise capital from various sources each source having a different interest cost. Details of the available projects and sources of finance are given in Table 2.1.

TABLE 2.1

Investment projects			Sources of finance			
Project	Outlay	Receipts one year later	Source	Amount raised	Required return	Repayment one year later
	£	£		£	%	£
1	100	115	C	100	10	110
2	100	111	B	100	8	108
3	100	107	A	100	6	106

Suppose it were decided to match projects and sources of finance in the way indicated in Table 2.1, i.e., Project 1 with Source C and so on. Project 1 would have a positive net present value of

$$£115/1·10 - £100 = £4·5$$

A similar calculation would show each project to have a positive net present value and on this basis all projects would appear to be acceptable. Reflection however will indicate that this would not be the best policy. Suppose Project 3 were not accepted. The organization would not then give up Source A but rather Source C, which is the most costly. In fact it would be better to reject Project 3 and to leave Source C untapped for the undertaking would thereby gain £3 (£110 − £107) in one year's time. (It is assumed that the supply of finance from C is not subject to a condition that it must be used for Project 1, a situation that may sometimes arise in practice.)

A better initial approach would have been to match projects and sources of finance according to a different scheme in which the most worthwhile project was matched with the least costly source of finance and so on.

According to this scheme Projects 1 and 2 would at once appear to be acceptable. Project 3 would not earn as much as the minimum return required by Source C and so should be rejected. The most expensive source of finance used now has a cost of 8% p.a. Each accepted project has a positive net present value when this cost of capital is used as a discount rate.

15

Although Example 2.3 is unrealistically simple it could be extended to deal with various practical complications without altering the essence of the conclusions. (In practice it would not usually be possible to ascertain precisely the cost of each additional slice of capital which had to be raised and it would be necessary to estimate the required discount rate before the full details of all the projects were known.) A point of principle is brought out by the example: that all worthwhile projects will have a positive net present value when their cash flows are discounted at the rate appropriate to the most expensive slice, or slices, of finance that will have to be used for a project, that is, the marginal cost of capital.

OPPORTUNITY COST

It was pointed out in Chapter 1 that since resources in general are limited it is seldom possible to undertake all projects which appear desirable in some absolute sense. Because of this, techniques of investment appraisal should be designed to ensure that scarce resources are used in a way which will yield the best possible return to the organization or community concerned as the case may require.

At the national level the government cannot obtain enough resources to build all the new roads, schools, hospitals, etc. that are desired by the community and it has to beware that it does not attract resources away from more worthwhile uses in private industry. Similarly an organization in the private sector may have several projects available, each of which would yield a return greater than that required to satisfy its shareholders, and yet find itself for the time being without sufficient finance to accept them all.

To be acceptable a project must pass a double test of worthiness. It must earn a minimum return equal to that required by investors on their money. In addition, it must earn at least as great a return as could be earned on any project which it displaces because finance is limited. If finance is scarce in this absolute sense a position of capital rationing is said to apply and the return which could be earned on the best project displaced by the one accepted can be called the internal opportunity cost of capital. The general principle is illustrated by Example 2.4.

16

EXAMPLE 2.4

Suppose, as in Example 2.2, that an organization is considering an investment project which requires an outlay of £10 000 and is expected to earn receipts of £4000 at the end of year 1, £5000 at the end of year 2, and £4000 at the end of year 3.   As before, the external cost of capital is taken at 10% p.a. this being the rate to be earned to satisfy the shareholders.   At this rate the project has a net present value of £773 as shown on page 13.

Since the net present value is positive the project can be said to satisfy the minimum criterion of earning a sufficient return to satisfy those who provide the finance.

Suppose, however, that there are alternative projects that would absorb all the finance available and could earn, it is estimated, 15% p.a. on money invested during the next three years.   If the first project is to earn 15% it must show a positive net present value when its cash flow is discounted at that rate.   The condition to be satisfied is that

$$-10\,000 + 4000/1{\cdot}15 + 5000/1{\cdot}15^2 + 4000/1{\cdot}15^3 \geqslant 0$$

In fact, however, it is negative, being -111.   The project should therefore be rejected in favour of one of the alternatives.

Hence it can be seen that an organization subject to capital rationing should only accept a project if it has a positive net present value when the internal opportunity cost of capital (15% in Example 2.4) is used as the discount rate.   If, however, there is ample finance available from external sources for all projects available the relevant opportunity cost of capital is the external cost, that is, it is the rate that has to be offered to attract or retain, as the case may be, external finance.

SHADOW PRICES

The internal opportunity cost of capital is an example of what is called a shadow price—here the shadow price of finance.   Any resource may have a shadow price if it is scarce in the sense of being absolutely limited in amount—so that it restricts an organization's ability to accept projects and a choice has to be made between different ones.   If more than one

17

resource is scarce, the calculation needed to achieve optimal selection of projects may involve the use of mathematical techniques (in which shadow prices are often known as dual prices) beyond the scope of this handbook.

ANNUAL EQUIVALENTS

In the foregoing examples the streams of costs and benefits arising at different times have been converted to equivalent present values so that they can be compared directly. An alternative method of comparison, mathematically equivalent, is to convert all the costs and benefits into constant annual equivalents. This is achieved by multiplying each present value by a specified factor, less than unity, which is a function of the rate of interest and of the relevant time period. Each item in the cash flow is then represented by a constant annual sum spread over the whole period of the project. An example of this approach is given in Chapter 5.

INTERNAL RATE OF RETURN

Instead of calculating the present value of a project at a stated discount rate the problem may be presented as that of determining the discount rate which would result in a zero net present value. This approach is usually referred to as the internal rate of return method.

The internal rate of return in Example 2.4 with a cash flow of $-£10\,000$, $+£4000$, $+£5000$, $+£4000$, is given by solving for S in the equation

$$-10\,000+4000/(1+S)+5000/(1+S)^2+4000/(1+S)^3 = 0 \qquad (1)$$

Solution of such an equation involves a process of approximation or trial and error since there is no method for direct solution. The only positive real value for S which satisfies equation (1) is, approximately, $0\cdot143$ or $14\cdot3\%$.

The criterion usually applied under this method is that a project should be deemed acceptable if it produces an internal rate of return greater than the minimum return required on capital. Thus, if the required cost of capital is 10%, the project should be accepted because the internal rate of return is shown to be greater than 10%. If the required return is set at 15% the project should be rejected. If the required rate of return differs from year to year this method cannot readily be used to reach a decision.

18

If it is merely necessary to decide whether a project meets a required standard rate of return it can be shown that (provided certain conditions are satisfied) the internal rate of return method must always lead to the same conclusion as the net present value method. If the required rate of return is r, constant over the life of the project, the net present value of the above project is

$$-10\,000+4000/(1+r)+5000/(1+r)^2+4000/(1+r)^3 \qquad (2)$$

A comparison of expression (2) with the left hand side of equation (1) shows that if $S>r$ expression (2) must be positive while if $S<r$ expression (2) must be negative. The same result holds in general provided S has a unique real value.

However, unless care is taken the present value and the internal rate of return methods of appraisal can lead to conflicting conclusions if two or more projects have to be ranked in order to ascertain which is the best.

EXAMPLE 2.5

Suppose Projects P and Q represent alternative ways of carrying out the same basic operation, that is to say, they are mutually exclusive projects. The problem is to decide which to accept. The cash flows in £s are as follows:

	Initial outlay	Cash receipts at the end of year		
		1	*2*	*3*
Project P	9000	3000	5000	6000
Project Q	9000	6000	4000	3000

If the cost of capital is 10%, the present values in £s are

Project P $\quad -9000+3000/1\cdot10+5000/1\cdot10^2+6000/1\cdot10^3$
$$= -9000+2727+4132+4508$$
$$= 2367$$

Project Q $\quad -9000+6000/1\cdot10+4000/1\cdot10^2+3000/1\cdot10^3$
$$= -9000+5454+3306+2254$$
$$= 2014$$

Acceptance of Project P adds significantly more to the value of the capital resources of shareholders than acceptance of Project Q and hence it may be decided to adopt the former. A comparison based on an internal rate of return calculation would have led to a different conclusion, as follows.

Project P $\quad -9000+3000/(1+S)+5000/(1+S)^2+6000/(1+S)^3=0$
whence $S=22\frac{1}{2}\%$ approximately
Project Q $\quad -9000+6000/(1+S)+4000/(1+S)^2+3000/(1+S)^3=0$
whence $S=24\%$ approximately
A comparison of the rates of return suggests that Project Q is the better.

The reason for the different results is that 6/13 of the total receipts of Q fall in year 1 and 10/13 in years 1 and 2, whereas only 3/14 of P's return come in year 1, and 8/14 in years 1 and 2. The initial investment is the same in both cases. The higher return in Q is thus earned on a smaller average investment and contributes less in absolute amount to the net value of the project. In some cases such a conflict can be removed by determining the internal rate of return on the differences between the two cash flows. Thus, in Example 2.5, deducting the cash flow of Q from P, the internal rate of return S is given by the equation:

$$-3000/(1+S)+1000/(1+S)^2+3000(1+S)^3 = 0$$
whence $S = 18\%$ approximately

demonstrating that P is preferable to Q. In other words, the greater amount of money value remaining invested in P at the end of year 1 (£3000) produces greater receipts in years 2 and 3. The differences in the receipts represent a return on the £3000 above the cost of capital of 10% p.a.

A further difficulty of the internal rate of return method may arise if a cash flow includes negative elements of significant size following positive ones, as may happen, for example, if there is a large payment to be made at the end of a project's life. In such cases the equation for the rate of return may have no positive real solution, or may have more than one.

Ways of adapting the internal rate of return (or yield) method to avoid these problems are discussed in MERRETT A. J. and SYKES A. *The finance and analysis of capital projects.* Longmans, 1963, Chapter 5.

SELECTION OF THE DISCOUNT RATE

Selection of the appropriate discount rate for an investment appraisal is of the greatest importance, since a wrong choice can lead to erroneous conclusions. For instance, the adoption of a high discount rate will reduce the present worth of future payments and receipts more than would the use of a lower rate. For this reason a comparison between the present worth of two cash flows which have different patterns over time will depend upon the discount rate used. If these patterns are significantly different, as they are in Example 5.5, the selected discount rate becomes a vital factor in the choice between two alternative projects. The choice will be determined by circumstances and requires the exercise of judgement concerning the future availability of finance and the opportunities for its use. In spite of the practical difficulties it is important to be clear on the principles which should guide selection.

The various sources of finance available fall into two main classes

    (a) fixed interest loans, including debentures
    (b) equity capital normally called ordinary share capital.

Money may also be obtained from preference shareholders who provide finance in return for a fixed annual dividend payable only if profits are made. For tax reasons preference shares are not at present an important source of finance.

The word interest in the context of class (a) refers to the periodic cash payments that must be made to a lender under a loan contract. It has the same connotation as the word dividend in relation to a shareholding. Interest in this sense must be distinguished from interest in the sense of a discount rate or rate of return which represents an estimate of the cost or sacrifice involved in investing in a specified project, and which is used in calculating the present value. The size of the rate of return may indeed be determined by contractual interest paid in the former sense. It may also however be determined by dividends expected by shareholders or by a rate of return on an alternative investment. Which of these is relevant depends upon the circumstances.

A special case of the use of a fixed interest loan is the temporary financing of a project by means of a bank overdraft. However, the use of a bank overdraft does not normally provide for the permanent needs of a project, and the interest rate payable on a bank overdraft may not offer a reliable guide to the cost of raising long-term capital.

Two kinds of situation may be considered in examining the appropriate discount rate which should be applied. The first is one in which funds are available for all projects; the second arises when the available finance is inadequate.

In the first situation although finance may be available its cost will nevertheless tend to increase with the amount needed and if the requirements of all projects are to be met some of the more costly sources may have to be utilized. It is seen from Example 2.3 that in such circumstances the rate employed for discounting should tend towards the rate attached to the most expensive source of finance actually used. This may involve a trial and error process of calculation.

In the second situation, illustrated in Example 2.4, in which the available finance is insufficient to meet all needs, competition for available supplies will arise. In these conditions the rate used for discounting should be higher than the external cost of finance and should correspond, or tend towards, the return expected from the best of the projects which has to be rejected through lack of capital, that is to say, it should represent the internal opportunity cost of capital. Here again a trial and error process of calculation may be needed.

If a project in the private sector is to be financed by the issue of equity capital the external cost of capital should correspond with the long-term return which the ordinary shareholders may reasonably expect from a company of the type concerned, having regard to all the circumstances. This will reflect the shareholder's view of the rate of return he could earn by investing his money in other companies.

If a project is financed by a mixture of loan and equity capital it is convenient to treat both types of finance as a single combined source, the external cost of capital being a weighted combination of the equity cost of capital and the cost of the loan capital.

Concerns in the private sector cannot raise all their finance on fixed interest terms. A substantial proportion of the capital provided has always to be in equity form. The equity shareholders are the first to suffer the effects of losses. In return they share in the whole of any residual gains.

In assessing the cost of obtaining finance, a borrower will take into account prevailing market rates of interest and rates of return on equity shares. Broadly, these reflect the demand for finance (which stems from investment opportunities in the market) and the supply of finance derived ultimately from the savings of individuals or of the government, or from overseas lenders. In a perfect market, in which lenders were completely certain of the return they would receive, there would be only one rate of interest at any time. In practice, the capital market is far from perfect and many different market rates of interest for finance of different kinds and different periods of time can be identified at any one time.

The government uses various devices to control the general level of interest rates as part of its general management of the economy. It also rations the amount of capital which may be used in various parts of the public sector, for example by the control of grants for public authority projects, by restricting the approval of ministries for the raising of loans by local authorities for the financing of capital projects, and by limiting the funds available to the Public Works Loans Board.

Another factor which powerfully influences the choice of interest rate is the degree of uncertainty which surrounds the outcome of a particular project. In general the greater the uncertainty, the greater will be the cost of capital.

The various sources of finance commonly employed by a business enterprise are described in Appendix B.

Ordinary shares in the private sector (called common shares in the United States and Canada) may be looked upon in some respects as analogous to finance provided by the taxpayer in the public sector. In the first case the shareholder invests his money in the hope of receiving in the future a regular stream of cash benefits of unspecified amounts. In the second case the taxpayer's money is employed to provide social or public services from which he may benefit as a member of the community.

Probably the best way of estimating the cost of new finance is to estimate the return currently being earned on the different classes of capital previously issued by the enterprise. Those from whom new finance is sought will be unwilling to accept a lower return than that offered by existing securities since the latter can be bought on the market. There is, however, a substantial degree of uncertainty, since the estimation of the current rate of return requires in principle the estimation of future dividend

3

flows and the solution of an equation of the same form as equation (1) on page 18.

Moreover, if a new issue of shares is large in relation to the capital previously issued, it may be thought that increased risks are associated with it and because of this it may be necessary to assume that shareholders will require a higher rate of return than in the past to justify the use of funds. Similarly the borrowing of a relatively large sum of money on fixed interest terms may involve the offering of a higher rate of interest. If no securities of the same class have been issued previously, some guidance may be found by studying the return offered by similar stocks issued by other organizations of similar standing but it is usually necessary to seek the advice of merchant bankers or other financial experts.

NEED TO FOLLOW FIRST PRINCIPLES

Various examples have been given in this chapter and certain rules have been derived from them. What is important, however, is the understanding of the principles on which these are based. No rule can fit all circumstances. It is essential in practical cases to be satisfied that the rules applied can be justified. The practising engineer must always be ready to return to first principles when the problem on hand requires it.

Chapter 3 Project cash flows

CASH FLOW OF A PROJECT

In the previous chapter it was shown how the net present value of a project can be calculated by discounting its cash flow. The cash flow of a project is specified by listing the expected cash receipts released by the project and the payments absorbed by it together with the dates on which each receipt or payment is expected to arise.

ACCOUNTING PROFIT

The cash flow of a project is a different concept from the accounting profit. This can be illustrated by taking a case where the project is a whole undertaking. In its most general form the profit of an undertaking is represented by the growth of its net present value as the result of its operations. Accounting profit as normally calculated is based on this idea, but the valuations used in its calculation are determined by standard procedures or rules of thumb and not by discounting cash flows. The accounting profit of an undertaking can therefore be defined as the difference between the values of the net assets (assets less liabilities) at the beginning and at the end of the period considered, taking into account the value of any assets which have been withdrawn or introduced by the owners, the values being determined on the basis of standard conventions. This may be seen from Example 3.1.

EXAMPLE 3.1

The figures represent the assets and liabilities of an undertaking as determined on the first and last days of the year.

	1 January	*31 December*
Land and buildings	£800	£800
Plant and machinery	200	0
Stock of materials	200	350
Debtors	540	720
Balance at bank	280	510
	2020	2380
Less creditors	370	340
Net assets	£1650	£2040

Let the owner of the undertaking have drawn £500 in cash from it during the year. The accounting profit calculation would then be:

Net assets, 31 December	£2040
Add cash withdrawn by owner	500
	2540
Less net assets, 1 January	1650
Profit	£890

Here the cash flow to the owner in the year was £500. The profit reported was £890. There is no precisely defined relation between profit and cash flow because the profit calculation in such a case is based, as already noted, on rule of thumb assessments designed mainly to satisfy various legal requirements; furthermore the cash flow of £500 covers only one year and therefore represents only one item in the time series.

An assessment of profit for the year based on an economic evaluation could also be made. This would involve first estimating the future cash flows to the owner (a) as seen at 1 January and (b) as seen at 31 December. These two cash flows would then be discounted, using the estimated cost of capital to the enterprise, to give net present values for the whole undertaking at the two dates. Suppose these were, respectively, £3000 and £4000. The profit on this basis would be taken, in £s, as

$$4000 - 3000 + 500 = 1500$$

(The £500 would again be added since it, or an estimate of its amount, would have been included in the cash flow discounted to 1 January but would not have formed part of the flow discounted to 31 December.)

Such a method of profit assessment can be regarded as a by-product of successive economic evaluations. It is not used for normal accounting profit calculations because of the high degree of subjectivity involved in its calculation, which makes it normally unsuitable for use where legal considerations are important.

Example 3.1 relates to a whole enterprise. The principles apply equally both to divisions or sections of enterprises, and to individual projects. In the latter case the cash flow takes the form of cash absorbed or released by the project at the expense, or for the benefit, of the rest of the enterprise.

The existence of accounting information prepared for purposes other than project appraisal sometimes leads to misunderstanding. The following points relate to particular areas of project appraisal which may cause difficulty.

(a) If acceptance of a project involves the use of an asset which is already owned by the undertaking the decision should not be affected by the value set on that asset in the balance sheet and books of account. This is frequently merely a historical record. If the asset would be sold were the project not undertaken it is its net selling value that should be included as a cash payment of the project since this is a receipt forgone in accepting the project. If, instead of being sold, the asset could be used in an alternative way that itself would produce a cash flow it is the latter that is forgone and must therefore be included as a cash outgoing of the project.

(b) Payments already made or for which firm commitments have been made before the decision should not be included in the cash flow for appraisal purposes. Suppose a feasibility study is carried out on a project and at the end of the study the project is appraised on the basis of all the information available. The cost of the study has already been incurred and cannot be affected by the current decision. It should therefore be ignored (unless the knowledge arising from it could be sold if the project were not undertaken, in which case it is the selling value that should be included). What has happened in the past is a bygone that cannot be changed. Only alternatives that lie in the future are relevant in making decisions.

(c) The cash payments and receipts to be included in a project's cash flow are the differences between the estimated total payments of an undertaking first if the project is carried out and second if it is not undertaken.

These differences may be called avoidable payments and receipts. This is true even if an accountant for some purposes would regard a proportion of some non-avoidable payments as a cost of the project (for example, part of the general administrative costs of the organization).

(d) If the acceptance of a project involves the acquisition of assets having long lives the payments for these assets are included as outlays in the cash flow of the project at the time or times of payment. No separate allowance is needed for depreciation in the normal accounting sense. Indeed, there would be double counting if annual depreciation were also included. Any residual cash value the assets have at the end of the project is taken in as a cash receipt at that time.

(e) In general, interest costs and the repayment of capital borrowed to finance the project should be excluded from the cash flow because, as was demonstrated in Example 2.1, the discounting calculations provide automatically for the payment of interest and the repayment of capital.

COSTS AND BENEFITS NOT MEASURABLE IN CASH

It has been assumed so far that the costs and benefits of a project arise in cash or are readily measurable in money terms. However, there will be many occasions in practice when this is not so. This is particularly likely to be true of a project in the public sector. There are, for example, difficulties in assessing the cash value of the benefits to be gained from building a public library or a nuclear submarine.

In spite of these difficulties, a discounted cash flow analysis can play an important role in decisions on such projects. Even if all the costs and benefits of a project cannot readily be reduced to money values it is still helpful to identify as many as possible. A discounting procedure is particularly useful in comparing different ways of carrying out a given project. For example, a stream of benefits not expressed in money values may be obtainable in different ways each giving rise to a different stream of payments. Other things being equal, the way associated with the cost stream having the lowest present value will be chosen. Again, it may be helpful to assess the present value of the flow of costs associated with a project in order to judge whether non-measurable benefits to be derived from it are worth the sacrifice.

NATURE OF INFLATION

Inflation is the process by which the purchasing power of money declines over time: in other words the amount of money required to purchase a given quantity of goods rises as time passes. If all consumers purchased an exactly similar collection of goods and services in each interval of time, the percentage increases in the sums of money successively required to make the purchases would provide a well-defined measure of the rate of inflation. In practice the measurement of the rate of inflation is less clear-cut because different people have different tastes and different incomes and hence buy different collections of goods and because in any period the costs of different goods tend to increase by different percentages. Hence, any measure of the rate of inflation has to be based on assumptions about the way in which consumers' money is spent. The problem is further complicated by the fact that the mix of any individual's purchases will change over time, partly because his income changes, partly because his tastes and needs change and partly because the goods available and their relative prices also change. Hence an index of inflation must be regarded as an approximation. An indicator is to be found in the *Monthly Digest of Statistics* prepared by the Central Statistical Office and published by HMSO. This Digest gives an index, related to a specified base year, of retail prices based upon an analysis of personal consumption.

CAUSES OF INFLATION

The total stock of money in the community may be divided into money available for spending and that which is not available because it is held by individuals, firms and others as a reserve for contingencies or as a working balance.

Consider the effect of an increase in the stock of money. If the average amount of money which people desire to hold in their pockets or in the bank remains unchanged the amount of money offered in exchange for goods must by definition increase and if the quantity of goods and services sold remains unchanged the increase in the flow of money can only be absorbed by an increase in prices. Thus an increase in the stock of money makes inflation possible. However, inflation will not necessarily follow. If, for example, the increase in the stock of money is taken up wholly by additional unspent holdings of money, no increase in the amount of money offered in exchange for goods will take place (assuming no change in the quantities of goods and services sold). On the other hand, some degree of inflation is possible even with no rise in the stock of money if there is a fall in the amount withheld from spending.

Even if the amount of money spent increased, an increase in the general price level would be avoided if the quantity of goods and services sold increased in like proportion.

If the productive resources of the community are substantially under-employed an increase in the amount of money spent will tend to call forth an increase in the flow of goods and services and there may be little or no increase in the general price level. If, on the other hand, productive resources are already near to full employment, little increase in the flow of goods and services will occur and prices will rise.

The foregoing discussion may be summarized by saying that there are theoretical and historical arguments for the view that a rate of increase in the stock of money held by a community not matched by a similar rate of increase in the level of productivity will, sooner or later, give rise to an increase in the general price level.

A common cause of an increase in the stock of money is the issue of additional money to finance government expenditure as a politically more acceptable course than increasing taxation, or than borrowing from the public at high interest rates. Inflation is particularly marked in war-time when government expenditure is running at a high level and resources are fully employed.

In a period of full employment and booming trade, the bargaining power of employees becomes particularly strong and they are well placed to force an increase in wage earnings without a corresponding increase in productivity; while employers, expecting high sales for their products and fearing loss of part of their labour force, have less incentive to resist the pressure. The resulting increased costs of manufactured goods are reflected in increased prices to consumers. This process cannot operate, however, unless the money flow can rise to a sufficiently high level to support the higher wage and price level.

While the foregoing explanation of some of the causes of inflation represents an over-simplification of an extremely complex problem, further discussion of the subject is not justified for the purposes of this handbook.

INFLATION AND PROJECT APPRAISAL

Inflation has to be considered in project appraisal both in estimating the cash flows of a project and in selecting the discount rate. In estimating the cash flows the main essential is that all estimates should be made on a consistent basis. All cash flows should be expressed either in pounds having a common purchasing power, for example in terms of purchasing power when the appraisal is actually made, or as the actual cash flows expected in the future after assessing the effects of inflation. Suppose that it is estimated that a sum assessed at £100 on the basis of current prices will have to be expended on a project in n years' time, and that future inflation is expected to average 4% p.a. The relevant item in the cash flow can be expressed as £100 if all estimates are in terms of constant purchasing power at present price levels or as £100 $(1 \cdot 04)^n$ if all estimates are to be in terms of the actual amounts of money paid or received.

It was stated earlier, as a basic principle, that the minimum rate of return required on an investment should be that rate which ensures that the investment will yield benefits at least as large as could be obtained from alternative uses of the finance. One of these alternative uses is personal consumption. In considering this alternative it is not sufficient to base the selection of a discount rate on indifference between, say, present consumption of £100 and consumption of £110 in one year's time without making it clear whether the latter amount of £110 is the actual sum of money required at the later date or the sum which would be required if prices of consumer goods were to remain constant. Suppose it is the former. Then the required return may be expressed as 10% per annum in actual money values in which case the cash flows of the project to be appraised should be the actual flows expected including inflationary increases. Alternatively, assuming inflation is expected to be at the rate of, say, 4% p.a. the £110 may be expressed as £110/1·04 or approximately £106 in pounds of purchasing power at current prices. The return is then expressed as 6% in terms of constant purchasing power or in 'real terms' as it is usually called and the cash flows will similarly be converted to their present-day equivalents in purchasing power.

The same argument applies if the alternative to the project is another investment. The project chosen should yield a return at least as great as any alternative investment of equal risk which may be available. The return required to be earned on such an investment, and the expected cash flow, may be expressed either in actual money terms or in real terms; the two forms of expression lead to the same investment decisions provided that they are applied consistently throughout the appraisal.

EXAMPLE 3.2

Consider again Example 2.2. The project required an initial outlay of £10 000 and promised cash receipts of £4000, £5000 and £4000 after 1, 2 and 3 years. Suppose that these cash receipts represent expectations expressed in terms of the actual cash to be received allowing for given expectations of inflation. Suppose further, that a minimum return of 10% p.a. in money (as distinct from real) terms is required. Discounting the cash flow at 10% p.a. the present value is found to be, in £s:

$$-10\,000 + 4000/1{\cdot}10 + 5000/1{\cdot}10^2 + 4000/1{\cdot}10^3$$

$$= 10\,000 + 3636 + 4132 + 3005$$
$$= 773$$

Suppose the expected rate of inflation is 4% p.a. The cash flow in £s expressed in real terms would be

	Initial outlay	Cash receipt at the end of year		
		1	2	3
(1) Money cash flow	10 000	4000	5000	4000
(2) Price index		1·00	1·04	1·04²	1·04³
(3) Real cash flow (Row 1/Row 2)	10 000	3846	4623	3556

If the money rate of return on an investment is required to be 10% and if the average rate of inflation is expected to be 4%, the required return in real terms will be approximately 6%. That is to say, the real rate of return required is approximately equal to the difference between the money rate of return and the rate of inflation. More accurately, if m is the required money rate of return and i is the rate of inflation, both compounded at the end of each year, the real rate of return is

$$\frac{1+m}{1+i} - 1$$

Discounting at 6% p.a. (the required return in real terms) the net present value of the project is found to be, in £s:

$$-10\,000 + 3846/1{\cdot}06 + 4623/1{\cdot}06^2 + 3556/1{\cdot}06^3$$
$$= -10\,000 + 3628 + 4114 + 2985$$
$$= 727$$

If it had not been for the element of approximation in the calculation the answer would have been identical with that obtained previously. In general it is usually more convenient to base estimates of future expenditure on the assumption that the current price level will continue, rather than to attempt making forecasts of inflation. Changes in prices of particular goods or services in relation to the general price level should however be allowed for.

TAXATION

In discussing various methods of investment appraisal the effect of such taxes as income tax and corporation tax has so far been ignored in order to simplify the exposition. In practice, careful attention must be given to this question. Special considerations may apply in the case of government projects (see page 35). In other cases the accepted rule is to treat a payment of taxation as a negative component of the cash flow, like any other payment. The discounting rate to be used must similarly be net of tax, that is to say, it should reflect the rate of return on the best alternative use of finance after taking into account the effect of taxation in that use.

It might at first seem that in the case of a percentage tax on income the same result would be obtained by ignoring the tax, provided that the discounting rate was also taken gross. However, this is fallacious. Taxation is assessed on the basis of a conventional accounting profit calculation and not on the net cash flow of a project. The accounting profit reported year by year will usually have a different pattern over time from that of the net cash flow. Moreover, the calculation of taxable profit is affected by statutory rules such that any given rate of tax does not necessarily fall on all components of the profit: the taxable profit is usually different from that reported to the owners of the enterprise.

The difference between (a) the time pattern of cash flow and (b) the time pattern of taxable accounting profits may also vary from project to project according to the incidence and amount of capital expenditure required for each. The effect of this variation is accentuated if under the taxation law some types of capital expenditure can be deducted from profits as depreciation at an earlier date than others. The earlier the tax relief from these deductions is received the greater is its discounted value to the project. For instance, especially favourable treatment may be given to expenditure incurred in establishing a new factory in a depressed area while encouragement of certain types of investment may be offered in the form of an investment grant towards the purchase of new assets. On the other hand, some payments included in the cash flow of a project may not be allowed as a deduction at all in computing taxable profit.

A detailed description of the United Kingdom rules for calculating taxable profit is beyond the scope of this handbook. It must, however, be emphasized that the effect of taxation can influence significantly the choice between different investments and the decision to carry out a particular project. For example, at the time of publication, a developer in Great Britain may be liable to pay tax under the following heads:

(a) local taxation (rates) levied by a local authority on the value of land and premises
(b) taxation levied by the central government on net accounting profits
(c) levy on the gain in the value of land as a result of development
(d) capital gains tax.

There are important differences between the positions of private firms and public authorities in regard to tax liability. For example, all private concerns have a general responsibility for all forms of taxation, while local authorities are liable to pay rates on certain classes of their assets including housing, but not on highways, sewers and public parks.

Nationalized industries are at present subject to both local taxation and central government taxation as if they were ordinary businesses.

United Kingdom taxation affects the cost and manner of raising finance for an enterprise. Interest paid to the holders of fixed interest securities is allowed as a deductable expense in computing the amount of profit on which taxation is assessed. Suppose that a loan stock is issued and repayable at par, and has an interest cost of 6% p.a., and that the enterprise is a company taxed at 40% on profits. The loan then has an effective net cost of only 3·6% since 2·4% is recovered in the form of taxation relief. Dividends paid to shareholders, however, are not allowable as a deduction in computing the taxation liability. In computing the weighted average external cost of capital of a company it is therefore necessary to combine the net of tax interest cost of fixed interest securities with the gross cost of ordinary share capital. The matter is further complicated because the present United Kingdom system of taxation makes it more worthwhile to finance an investment from retained earnings (funds belonging in principle to the ordinary shareholders but not paid out to them, see Appendix B) than from a new issue of ordinary shares.

All project appraisal involves the estimation of future cash flows. Taxation forms an important negative element of these. The estimation of future tax rates can therefore be an essential part of project appraisal. Because tax rates depend upon the action of government, this estimation is subject to an especially high degree of uncertainty. Nevertheless, unless for special reasons it is decided to ignore tax in an appraisal, the attempt has to be made. Where there is uncertainty the engineer should seek specialist advice.

When projects are carried out by or for the central government of a country, general economic considerations may indicate that certain of the taxes that would otherwise be payable should be ignored in the economic appraisal. For example, the appraisal of a project in an underdeveloped area may be made on the assumption that no duties will be charged on expensive equipment that will have to be imported.

Such cost–benefit decisions involve judgement and, in the end, are matters to be settled at the political level on the basis of expert economic advice. However, the rules for certain types of decision (such as the suspension of import duties referred to above) may be sufficiently well-established for the engineer to assume that he can follow them in the absence of special instructions. Normally this will be when he is familiar with practice of the type concerned. For reading on the economic principles underlying decisions on these matters, and on those discussed in the first section of Chapter 4, see: PREST A. R. and TURVEY R., Cost–benefit analysis: a survey. *Surveys of Economic Theory*. 1967, Macmillan.

Chapter 4   Project appraisal in the public sector

SPECIAL CHARACTERISTICS OF THE PUBLIC SECTOR

Projects in the public sector of the economy comprise those undertaken by the nationalized industries and by central and local government.   The principles of economic evaluation are basically the same as in the private sector.   Appraisal requires the estimation of the flow of costs and benefits associated with the project and the choice of a discount rate.   The object is to determine whether a particular project or method of carrying out a project, judged from the point of view of the community as a whole, will result in net benefits of greater value than could be obtained from any alternative use of resources which has to be forgone if the project is undertaken.

If all the costs and benefits of a public investment can be expressed as a cash flow, and a required return is expressed as a discount rate, the project can be evaluated by a net present value calculation of the type already discussed.

However it is often difficult to attach a money value to all costs and benefits. In particular, the benefits are often not of a kind which can be bought and sold in any market.   In recent years economists have devoted considerable effort to the development of methods for assessing the monetary value of the costs and benefits arising from public expenditure.   The subject, known as cost–benefit analysis, is complex.   It must be admitted that for certain types of public expenditure (such as defence expenditure) it is hardly possible to derive an acceptable method of evaluation.   In such circumstances the justification for the proposed project must be based on judgement in the light of the best information available.   Even in such cases, the study of the comparative costs of different ways of achieving a given objective may be useful and important; and an assessment of the identifiable costs is an important element in providing the basis of a political decision.

Difficulty also arises in the choice of a discount rate. A study of the investment market will provide guidance as to the rate of return which a private investor is able to obtain in a defined set of circumstances; but to the extent that public investment is financed out of taxation, the element of personal choice is absent and there is no corresponding standard of reference. Even where specific loan finance is used for a project the special considerations relating to government finance make the rate of interest paid on it an unsure basis for the discount rate for project choice. A loan to the government is looked upon as very secure because the whole taxing capacity of the State is available as a guarantee of repayment. Loans raised by local authorities and by nationalized industries are in a similar position because they too are usually supported by substantial taxing capacity or by the guarantee of the central government.

However, this does not justify investment of borrowed money on the basis of an expected return which just covers the interest cost. Regard should be given to the return on other public or private investments available to the community. Furthermore, the expected return should be related to the uncertainty of the project. Generally, the return sought on a public investment financed by borrowing should be the same as that which would be required if the same investment were financed by taxation and should be determined on the basis of the return that could be obtained on the best alternative use of the finance in the economy as a whole.

If money were not taken from an individual in the form of taxation he could either consume it or invest it in the private sector of the economy where it would earn for him a rate of return compatible with the risks involved. It can be argued that the community as a whole should not look with favour upon an investment in the public sector unless it is believed that, for equal risks, it is likely to yield as great a return as could be hoped for in the private sector. Thus, it can be said that the best returns that would have to be sacrificed in the private sector in order to allow investment to occur in the public sector provide a guide to the discount rate which should be used for appraisals in the latter.

CENTRAL GOVERNMENT FINANCE

The sources of finance for investment in the public sector fall into two groups, internal and external.

Internal sources of finance are mainly confined to nationalized industries. In these industries the internal cash flows provide a source of finance for the industry's investment. In effect this source forms part of the annual receipts of the central government, since it reduces the amount of external finance that has to be provided by the government for the industry's growth. External sources of finance are taxes and borrowing.

LOCAL AUTHORITY FINANCE

In the case of local authorities money may be provided by a grant from the central government funds or may be raised by the local authority's own borrowing or taxation (rates). In general, local authorities meet current expenses (the costs of day-to-day administration, the general maintenance of public services, and interest on money they have borrowed) out of the rate fund raised by local taxation which may be augmented by the profits of successful local authority enterprise, if any.

Government grants are made for current expenditure (notably to assist in defraying the cost of education) and may also be made for certain types of local capital expenditure.

Local authorities borrow money for capital investment by issuing securities in the ordinary market at the current rate of interest. They may also borrow from the Public Works Loan Board. This is a government agency for providing access to capital. It is provided with money from central government funds and charges a current rate of interest on its loans. Local authorities also accept money on deposit at interest.

The usual practice is to raise loans from time to time as may be necessary to meet all financial needs, including repayment of earlier loans, and not to associate a particular loan with a particular investment project. From this practice arises the concept of a consolidated interest rate which represents a weighted average rate notionally attached to the whole of the outstanding debt at a particular date. It is calculated from the different rates of interest at which money has been borrowed over a long period of years. This consolidated interest rate is sometimes used when preparing a budget or financial plan for a project for the purpose of planning the repayment and interest outlays which are likely to arise.

For reasons given in Chapter 2, a rate of interest used for such a purpose is not necessarily equal to the rate which would be used for discounting purposes when making an economic appraisal of a project.   For instance, in Example 5.5 a discounting procedure is used to evaluate the relative costs of two alternative methods of providing an electric power supply. Example 8.1 illustrates the method of funding a loan which has been used to finance the project.   It will be seen that the loan terms are fixed by the lenders with reference to the prevailing money market conditions. The rate used for discounting may, however, be higher if it is determined from consideration of the return which alternative projects are considered to be capable of earning.

THE NATIONALIZED INDUSTRIES

To the extent that their cash receipts exceed their cash payments, the nationalized industries generate their own internal finance.   In some instances this excess or part of it may be paid over to the central government as a kind of dividend over and above the interest that is paid on loans from government funds.   So far as such an excess is retained for investment or re-investment inside the industry, the central government has to provide correspondingly less external finance in the form of loans. Any excess of receipts over payments is therefore in effect a contribution to central government financing whether it is paid into the central funds or not.   The more profitable the industry the greater will be the amount of internal financing possible.   In this respect the more profitable nationalized industries resemble many companies in the private sector of the economy which generate a large proportion of the finance they use for expansion.

As regards external financing there is a marked difference between the two sectors.   In the private sector, for the reasons explained in Chapter 2, only a proportion of the capital is in the form of loans carrying a fixed rate of interest and some companies do not borrow at all.   The remainder of the capital, usually more than half, is in the form of equity shares (ordinary shares) which earn a variable rate of dividend depending upon the profitability of the undertaking and the dividend distribution policy of the board of directors.   In the nationalized industries, however, the whole of the external finance required is normally provided from government sources

4

in the form of fixed-interest loans. A fast-growing nationalized industry may be continually receiving fresh loans from central government each of which may have a different rate of interest attached to it. The money from these loans will form part of the general funds of the industry, available for all purposes. The industry is expected to pay the interest on the loans but on occasion this may have to be waived (as has happened in the case of British Railways).

As in the case of other types of public investment, it is not normally appropriate to assess new projects on the basis of the rates actually paid on loans. Such a course would be unlikely to reflect the opportunity cost of finance to the community and the government therefore lays down from time to time a discount rate to be used when assessing new projects (see for example *Nationalized Industries: A Review of Economic and Financial Objectives.* HMSO, 1967).

The nationalized industries are thus required to apply discounting tests in assessing new capital projects. Independently of these they are also set overall financial objectives. Over the long run these targets imply the meeting of all outlays, including interest actually paid on money borrowed, plus some specified surplus or minus some specified loss. The financial objectives vary from industry to industry.

COMMERCIAL CONTROL

The public and private sectors of industry differ in the degree of governmental control over their commercial operations, including price fixing. In the private sector competition usually operates as a curb on prices and a spur to efficiency and the amount of governmental regulation is relatively small. Some of this regulation is directed towards the preservation of competition and the avoidance of undue monopoly power.

The nationalized industries on the other hand have been granted monopoly privileges. By their nature they operate under the general direction of the government though they are given a great deal of freedom in their day-to-day operations. They usually have, for example, some latitude in their price-fixing though some guidance has been given on the general principles of pricing. These principles are, broadly, that customers and consumers of services should be charged prices which reflect the economic costs of providing the particular goods and services in question.

40

Chapter 5   Problems in economic choice

INTRODUCTION

An economic evaluation may be considered under a series of phases:

Phase (i)   define the problem to which a solution is sought

Phase (ii) set out the available data, state the assumptions made and assess the reliability which should be attached to them

Phase (iii) make the calculations

Phase (iv) interpret the results of the calculations and where necessary modify them on the basis of judgement to allow for factors not taken into account in the formulae used.

The definition of the problem usually implies that associated decisions have already been taken. For instance, the calculation of the optimum amount of reinforcing steel in a concrete bridge beam assumes that a series of more fundamental economic decisions has been made. Such decisions might relate to the use of a beam in preference to a slab, of concrete instead of steel, of a bridge instead of a ferry and so on. Thus the particular problem to be solved can usually be regarded as a step in a long chain of interconnected decisions. In principle all the decisions in the chain should be taken concurrently, since they may interact. In practice a step-by-step method may have to be used because the complexity and difficulty of planning for overall optimization may involve costs which exceed any likely savings. If there are strong interactions between the various steps they may have to be repeated a number of times.

The second phase concerns the selection of the data to be used and the assessment of their reliability. Bound up with this are the assumptions that must be made in applying the data to the solution of the problem and to the forecast of future events.   Upon the skill used in the selection of data depends the value to be attached to the results of the calculations.   Only by the exercise of judgement in relation to data and assumptions, and by an appreciation of their limitations, can the results of an economic analysis become meaningful and an assessment be made of the reliability of any conclusions reached.

The methods used in applying the selected data constitute the third phase. The methods are based on the principles explained in previous chapters. The examples which follow later have been chosen to illustrate a variety of applications.

In the fourth phase the meaning to be attached to the results of the calculations must be interpreted in the light of the engineer's knowledge of the particular circumstances which surround the problem and should have regard to factors that are not adequately dealt with by the formulae used in phase (iii).

Finally, it must be recognized that the validity of any conclusions drawn will depend on:

(a) the completeness of the items included in the stream of cash payments and receipts, i.e. the costs and benefits
(b) the reliability of the figures used.

EXAMPLES

In the following pages examples are given to illustrate typical problems in economic choice.   In most of these examples the economic evaluations have been carried out by calculating present values.   As explained in Chapter 2 an alternative method, that of annuitizing costs and benefits (expressing them as constant annual equivalents), is sometimes used.   The two methods are arithmetically equivalent and comparisons based on them yield the same answer.   In the majority of cases the present value method is easier to apply.   The annuity method is sometimes convenient when it is required to show the surplus or loss arising from a particular project in annual terms, for example, as a guide in fixing price tariffs.

In the examples all the data used are assumed to be in real terms, that is, the effects of inflation have been removed as explained in Chapter 3.

EXAMPLE 5.1

This example illustrates the equivalence of the present value and annuity methods of calculation.

A building can be lined with heat-insulating material at a cost of £16 500 which will reduce heating costs by £2600 per annum. It is assumed that the building has a useful life of 10 years at the end of which the insulating material has no residual value. The problem is to determine whether the proposal is worthwhile if the required return on capital is 8% per annum.

Present value (capitalized cost) method

The heating savings can be expressed as a present value by using Table 3 of Appendix F.

Present value of heating savings £2600 × 6·7101	£17 446
Initial cost of insulation	£16 500
	————
Net present value	£946

Hence, on the assumptions made, the project is worthwhile.

Annuity (annual cost) method

The initial cost can be expressed as the equivalent of a constant amount payable at the end of each of ten years. This is the amount which would just suffice to recover the initial cost together with interest on the outstanding balance from year to year. It is therefore the minimum annual saving needed to justify the initial expenditure on the basis of the above assumptions and on the assumption that this minimum saving is the same in each year. Let the required amount be £x. The initial cost of £16 500 is then the present value of an annuity of £x for 10 years at 8% per annum. Using Table 3 of Appendix F

$$x = 16\ 500 \times \frac{1}{6·7101}$$
$$= 16\ 500 \times 0·149\ 03$$
$$= 2459$$

The evaluation proceeds as follows.

Annual heating savings	£2600
Annual equivalent of first cost	£2459
	————
Net annual value	£141

Thus again the project is shown to be worthwhile.

43

The two methods must give the same answer since the figures used differ only by a constant factor. This can be shown by calculating the present value of the annual saving:

£141 × 6·7101 = £946

It will be observed that on the basis of the assumptions used in this example the margin of advantage in favour of insulating the building is small. In such circumstances further examination of the assumptions should be undertaken on the lines indicated in Example 6.2 where for illustrative purposes the same numerical values have been used.

EXAMPLE 5.2

This example illustrates the method of determining minimum net cost when this is made up of two mutually interdependent variables.

The problem chosen is the determination of the most economic area of copper conductor to carry a given electric current. The total cost $C = C_1 + C_2$ where C_1 is the initial cost of purchase and installation and C_2 is the present value, at the given rate of return, of the electricity dissipated in losses. The cost of purchase and installation C_1 is dominated by the cost of copper which will be proportional to its weight and consequently to the area of the conductor. Allowing a fixed cost F for installation, C_1 may be written:

$$C_1 = KA + F$$

where A is the area of conductor and K is a constant.

The cost of electricity dissipated in losses C_2 is dependent upon the resistance of the conductor which is inversely proportional to its area. This cost can be represented by the equation

$$C_2 = k/A$$

where k is another constant.

Hence the total cost $C = C_1 + C_2$ may be written:

$$C = KA + F + k/A$$

Differentiating with respect to A the minimum value for C will occur when

$$\frac{dC}{dA} = 0, \text{ or when } K - \frac{k}{A^2} = 0$$

whence $A = \sqrt{(k/K)}$

This expression can also be written $KA = k/A$.

Thus it is seen that the minimum cost occurs when the variable part of the initial investment cost equals the cost of the losses, provided they are both expressed on the same basis. In the foregoing solution the present value or capitalized cost method has been employed. It would have been equally convenient to use the annuity or annual cost method.

The solution to the foregoing problem can also be expressed graphically, the diagram being very similar to that used in Fig. 5.1 to illustrate the example which follows. In the case under discussion, the conductor area would be plotted along the base, the cost of the losses would be represented by a falling curve, and the initial cost of the conductor would be a rising curve (in this case a straight line). The lowest point on the total cost curve given by the vertical sum of the two curves would correspond to the conductor area which gives minimum cost.

EXAMPLE 5.3

This is a more complex example of the theme illustrated in Example 5.2 and deals with the determination of minimum total cost by examining the inter-relationship between the costs and benefits arising from a project.

An area of low-lying agricultural land is subject to frequent flooding which prevents its effective use. By building an embankment this flooding can be reduced or virtually eliminated depending upon the height chosen. The problem is to determine the optimum height of the embankment by considering the relationship between benefits and costs. The optimum height of the embankment will be reached when the present value of the expected surplus of benefits over costs is a maximum.

The primary benefits arising from the project will consist of a net increase in the value of the crops produced as a result of the protection afforded by the embankment. In practice secondary benefits are likely to arise since the land protected may become more intensively settled and may be put to other uses than agriculture.

An assessment of the benefits should take into account the effect of future economic growth in the area both with and without the project. Such an assessment should not overlook the possibility that the project may have adverse effects on economic growth in more distant areas.

The costs associated with the project are the capital cost of construction and subsequently the recurring cost of maintenance. The selection of the optimum height will depend upon these costs in relation to the estimated benefits arising from reduction in flooding. The costs will also depend on the flood level frequency relationship, the selected discount rate, and the life assumed for the embankment (the amortization period).

Taking these in turn:

(a) the costs of construction and maintenance of the embankment for various heights can be assessed from engineering studies

(b) the probability of the occurrence of floods of any defined level can be estimated from past records, proper attention being given to the possibility of future changes in régime as the result of the construction of the works

(c) the estimated annual benefits must take into account not only the reduction in crop losses which would arise from the reduced flooding if applied to the existing crop pattern of cultivation, but also future increased productivity as the result of improved agricultural methods made possible by the works. Such estimates are necessarily speculative and are likely to be subject to a considerable margin of error

(d) the discount rate selected will be that considered appropriate at the time the project is put in hand. It will be determined having regard to considerations outlined in Chapters 2 to 4

(e) the life assumed for the embankment will usually be long enough to make the total cost of the project little different from that based on the assumption that it will last for ever. (See Appendix E.)

A simple analytical solution to the problem of optimum embankment height follows. It is based on the assumption that the losses which arise each time the river overtops the embankment are independent of the duration and the peak level reached by the flood.

The following further assumptions have been made.

Loss from each flood which overtops the
embankment   £200 000
Life of embankment   50 years
Required rate of return   10% per annum

It is also assumed that the only net benefit of the project will be that arising from the avoidance of losses due to flooding.

Table 5.1 gives the relationship between embankment height and total cost, based on engineering studies in conjunction with the above assumptions as to discount rate and life of the embankment.

TABLE 5.1

Embankment crest level above datum m	Present worth of embankment and maintenance costs £
11	200 000
11·5	267 000
12	356 000
12·5	475 000

The frequency of water levels of 11 m or more is given in Table 5.2.

TABLE 5.2

1 Water level m	2 Expectation that peak annual water level will exceed the value in column 1 %	3 Return period m* years
11	25	4
11·5	10	10
12	4	25
12·5	1	100

* The return period m is derived from column 2 and equals $100/E$ where E is the expectation per cent.

Let A be the present value of future losses due to flooding and B be the present value of the cost of protective works. Then the surplus of benefits over costs will be a maximum when $(A+B)$, the sum of the losses and costs, is a minimum. A, the present value of future losses for a return period of m years between successive overtoppings, can be found approximately from:

$$A = \frac{L}{m} \times \left[\frac{1-(1+r)^{-n}}{r}\right]$$

where L is loss from each overtopping (£); m is the return period (years); n is the embankment life (years); r is the cost of capital expressed as an annual interest rate.

This expression involves the assumption that a loss, equal to the average annual loss, arises each year. The actual losses will be variable.

When n is large

$$A = \frac{L}{m} \times \frac{1}{r} = \frac{L}{mr}$$

Thus $A = \dfrac{\pounds 200\,000}{m \times 0.10} = \dfrac{\pounds 2\,000\,000}{m}$

and for different return periods has the values shown in Table 5.3.

TABLE 5.3

Water level m	Return period m years	A £
11	4	500 000
11·5	10	200 000
12	25	80 000
12·5	100	20 000

B can be read directly from Table 5.1. The values of $A+B$ are given in Table 5.4. They are plotted in Fig. 5.1.

TABLE 5.4

Embankment crest level m	Losses A £	Costs B £	A+B £
11	500 000	200 000	700 000
11·5	200 000	267 000	467 000
12	80 000	356 000	436 000
12·5	20 000	475 000	495 000

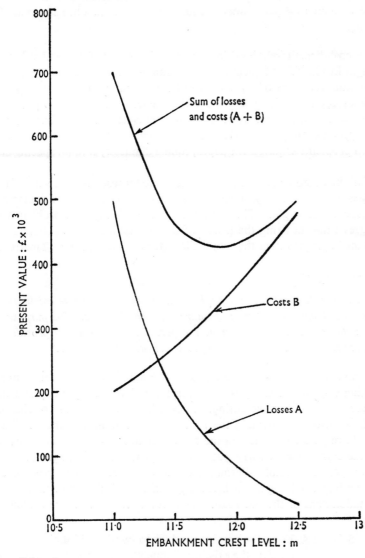

FIG. 5.1

It will be seen from Fig. 5.1 that the economic crest level is about 11·85 m above datum. This will reduce the frequency of flooding from an average of once in 4 years under natural conditions to once in about 12 years.

The cost–benefit relationship expressed in terms of the embankment height in Fig. 5.1 is typical of a large number of problems in which the optimum economy of a project can be assessed from a range of possibilities represented by a principal variable. Although the principal variable in this example is the embankment crest level, a similar method of approach may be applied in determining optimum dimensions in such problems as pavement depth, tunnel diameter, column spacing etc.

It will be noted from Fig. 5.1 that the curve representing the sum of the losses and costs is asymmetrical with regard to the main variable: the embankment height. This is characteristic of this type of problem and suggests that in fixing the height of the embankment it is better to err on the high side rather than on the low (for example, accept 12 metres).

EXAMPLE 5.4

This example demonstrates the application of net present value calculations to a project involving a stream of future costs and benefits. The purpose of the calculations is to show that over a period the benefits will provide an adequate return on expenditure, and to select the better scheme.

A gravel surfaced road on a reasonably good alignment through open rolling savannah country is now carrying 150 vehicles per day and the amount of traffic is increasing. A decision is required as to whether it is worthwhile providing the road with a bituminous surface. This would require strengthening the road base and it would be opportune at the same time to carry out minor improvements in alignment and drainage. A possible alternative would be to reconstruct the road on a new improved alignment involving a saving of 10% in distance, and abandonment of the old road. Thus the alternatives to be evaluated are:

> Scheme A bring the present road up to two-lane bituminous standard on the existing alignment with only minor improvements to horizontal and vertical alignment

Scheme B reconstruct to two-lane bituminous standard with major re-alignment which will lead to a saving of 10% in distance.

Costs and benefits

The cost elements to be considered are: the capital costs of the improvement or reconstruction; the recurrent costs of road maintenance.

The overall benefits of road improvement normally include: reduction in vehicle operating costs; reduction in accidents; time saved by users; new development associated with generated traffic.

There may be circumstances in which less tangible benefits such as better access to schools and hospitals and more comfortable travel have also to be considered.

The evaluation of such benefits often involves a complex mixture of calculation and judgement usually calling for co-operation with other specialists, and an adequate treatment is beyond the scope of this handbook. Decisions in the end are frequently made at the political level; the engineer's concern is to present the issues, showing the results of the cost–benefit calculation, and drawing attention where appropriate to other factors to which it may be difficult to attach values but which should nevertheless be considered.

In this example the effects on road accidents are likely to be small and have been neglected in the calculation. Time saved by road users may be significant, in which case a problem of valuing the time saved arises, and it will usually be necessary to consult local planners. Speeds on the bituminous road will be higher than on the gravel road, and on the re-aligned road the distance will be shorter. In this example the only allowance for time saved is that already counted in the estimate of reduced vehicle operating costs.

The reduced vehicle operating costs are likely to generate new development. Appraisal of these benefits is usually complex, involving co-operation with agricultural specialists, economists and other planners.

51

They are particularly important in development plans for regional road improvement and in the special case of improving a dry-season earth track to an all-weather road. In this example they could well be a significant factor and the engineer would normally consult with local planners to work out how these benefits should be evaluated. However, for simplicity in the calculations which follow no allowance has been included for such benefits. The omission of these and other benefits mentioned above does not affect the method of calculation. Once the benefits have been evaluated and expressed in the form of an annual series of cash equivalents they are treated in exactly the same way as the benefits assumed in the example.

The benefits counted here are confined to those accruing from reduced vehicle operating costs. Such costs on an earth or gravel road normally range between 1·1 and 2·0 times the costs on a bituminous road. In this example the existing gravel road has been assumed to be fairly good and the ratio of costs has been taken as 1·2:1.

Data

Some data (such as those just mentioned on the state of the existing road) will be obtained from direct observation and measurement in local conditions, for example, rates of traffic growth and vehicle operating costs. Both are likely to require judgement and where there is uncertainty it may be desirable to set confidence limits. In dealing with traffic growth, for instance, it may be desirable to present three sets of calculations, one based on the minimum estimate of traffic growth, one on the probable and one on the maximum.

It is assumed that the flow of costs and benefits arising from the investment will extend over a period of 20 years from the start of construction. Expenditure on construction is spread uniformly over the first 2 years at the end of which the benefits from reduced operating costs commence to accrue. Present values have been calculated using an 8% discount rate. Any net benefits arising after the end of the 20 year period are ignored.

The data assumed in this example are given in Table 5.5.

TABLE 5.5

Road construction	Length of existing road	30 km
Scheme A	Cost of improvement on present alignment	£240 000
Scheme B	Length of re-aligned road	27 km
	Cost of reconstruction of re-aligned road	£324 000
Road maintenance	Costs of maintaining existing gravel road per km per year	$£30 + £1 \cdot 24Q$ where Q is the average daily traffic flow (vehicle/day)*
	Costs of maintaining improved road under Scheme A per km per year	£200
	Costs of maintaining new alignment under Scheme B per km per year	£200
Road traffic It is assumed that the composition of vehicle flow will remain unchanged	Present flow Rate of increase	150 vehicle/day 10% p.a. (compound)
Vehicle operating costs These exclude duty and tax	Mean operating cost on existing gravel road	0·0415 £/km day
	Mean operating cost on two-lane bituminous road	0·0345 £/km day

* The formula for calculating gravel road maintenance costs is made up of a fixed charge (£30) for maintenance of verges, ditches, etc., and a variable element (£1·24Q) related to road usage. In practice it becomes virtually impossible to maintain a gravel road when average daily traffic reaches a level of about 350 vehicle/day. In the above example traffic at the end of 20 years is estimated at just over 900 vehicle/day, but in practice the assumption could be made that funds allocated to road maintenance would not exceed $30 + (1\cdot24 \times 350)$ £/km year but the road thereafter would deteriorate and the deterioration would be passed on in higher vehicle operating costs, at the same absolute value as the hypothetical increase in road maintenance costs.

THE CALCULATION

Vehicle operating costs

Let q be the annual rate of growth of traffic; r the annual rate of discount; and V the total annual vehicle operating costs in year 0 (preceding commencement of reconstruction).

53

Then the annual operating costs for n years form the series
$$V(1+q),\ V(1+q)^2,\ \ldots, V(1+q)^n$$

The present value of this series, discounted to the beginning of year 1 is

$$V\left[\frac{1+q}{1+r}+\left(\frac{1+q}{1+r}\right)^2+\ldots+\left(\frac{1+q}{1+r}\right)^n\right]$$

or

$$V\sum_1^n\left(\frac{1+q}{1+r}\right)^n$$

Scheme A

For years 1 and 2 the existing conditions hold. For years 3 to 20 the costs are reduced in the proportion of 0·0345 to 0·0415. The present value at the beginning of year 1 of the costs for years 3 to 20 is found by first calculating the present values for the years 1 to 2 and for the years 1 to 20 using the above formula, and then deducting the former from the latter. On the basis of existing conditions the present value is therefore

$$V\left[\sum_1^{20}\left(\frac{1+q}{1+r}\right)^{20}-\sum_1^2\left(\frac{1+q}{1+r}\right)^2\right]$$

The answer is reduced in the ratio of 0·0345 to 0·0415 since operating costs fall by this ratio from the beginning of year 3. Inserting values as follows

$$V = £0·0415 \times 150 \times 30 \times 365 = £68\,200 \text{ to nearest } £100 \text{ per year}$$
$$q = 0·10$$
$$r = 0·08$$

the present value is found as £1 267 000.

Scheme B

The present value for Scheme B, calculated on similar lines with additional savings in the ratio 27/30, is found to be £1 140 000.

54

Savings

The present value of existing costs for years 3 to 20 is found similarly to be £1 521 000.

Savings on vehicle operating costs are therefore:

Scheme A £1 521 000 − £1 267 000 = £254 000

Scheme B £1 521 000 − £1 140 000 = £381 000

Maintenance costs

Scheme A

After the first two years the annual costs per km remain constant at £200, or £6000 per year for a 30 km road.

The present value of £1 per year for 20 years, less the present value of £1 per annum for two years at 8% per annum is found from Table 3 of Appendix F to be:

9·8181 − 1·7833 = 8·0348

Multiplying by £6000, the present value is found to be £48 000.

Scheme B

Similarly the present value of the costs for Scheme B over a 27 km road is found to be £43 000.

Savings

The present value of the costs which would be incurred over the same period, that is for years 3 to 20 if nothing were done to improve the existing road, can be found by applying a formula, similar to that used for operating costs, to the data given in the footnote to Table 5.5. Such a calculation shows the present value to be £132 000.

Since the costs incurred during the 2 years construction period are common to both Schemes A and B they may be neglected. Thus the savings on maintenance costs may be expressed as follows.

Scheme A £132 000 − £48 000 = £84 000

Scheme B £132 000 − £43 000 = £89 000

The results may now be summarized.

Scheme A
    (a) Present value of savings in vehicle operating costs £254 000
    (b) Present value of savings in maintenance costs £84 000
    (c) Present value of construction costs £214 000

The net present value of Scheme A is (a)+(b)−(c)
$$= £254\,000 + £84\,000 - £214\,000 = £124\,000$$

Scheme B
    (a) Present value of savings in vehicle operating costs £381 000
    (b) Present value of savings in maintenance costs £89 000
    (c) Present value of construction costs £289 000

The net present value of Scheme B is (a)+(b)−(c)
$$= £381\,000 + £89\,000 - £289\,000 = £181\,000$$

Appraisal of results
On the assumptions made both Schemes are profitable since each shows a positive net present value. As Scheme B, involving a major re-alignment of the existing road, shows a larger net present value than Scheme A it would be preferred. In practice it would be desirable to test the effect on the net present values of altering the assumptions, for example, with respect to growth and mix of traffic flow.

EXAMPLE 5.5
This example again demonstrates the use of net present value calculations as a means of making a selection between two alternatives which differ significantly in their relationship between initial (capital) costs and running costs. As a consequence of these differences the choice is sensitive to changes in the discount rate.

The problem presented is the choice between hydro and thermal generation as a means of meeting a growing demand for electricity. In the example which follows it is assumed that the market for power is the same for both projects, that is, there is no difference between their benefit flows and hence they may be compared by considering only their respective cost flows. It is further assumed that there is no difference between the costs of electric transmission associated with the two projects.

56

The more important differences which characterize such projects may be summarized as follows.

Hydro	Thermal
(a) Higher capital cost and lower running cost	(a) Lower capital cost and higher running cost
(b) Longer life of assets	(b) Shorter life of assets
(c) Basic engineering studies are more complex and cost estimates less certain	(c) Cost estimates less subject to error
(d) Costs of hydro plants built in the future not expected to fall greatly as the result of technological change	(d) Costs of thermal plants built in the future sensitive to and likely to fall as the result of technological advance
(e) A large part of the total capital expenditure for present and future needs has to be incurred initially	(e) Much of the expenditure required to extend the power station is incurred as required to meet demand thus providing some flexibility in planning.

Some of these differences may be seen from the data given in Table 5.6.

Before a valid comparison can be made between two projects with such widely different characteristics, the engineering investigations must be carried to a stage which ensures the adoption of the most appropriate and efficient technical solution for whichever project is selected. This is especially important with hydro power since the cost is so much dependent upon topography and other purely local conditions. Thus a large part of the work involved in economic evaluation lies in the engineering studies which provide the basis for its formulation.

A fundamental purpose of such an evaluation is to select the method of generation which will lead to the provision of electricity at the minimum economic cost, taking into account the load pattern and the standard of service required. Experience shows that if both sources of power are available this objective can usually be achieved by their combination rather than by treating them strictly as alternatives. A consequence of such a combination is that the service provided by the hydro project tends to be somewhat different from that which would be provided by an alternative thermal project. For example, the average annual load factor at which a hydro project will operate will normally remain unchanged throughout its life, while that of a thermal project will decline with age. A rigorous economic study of the problem requires that the evaluation should cover an extended period of years, should examine the effect of alternative programmes and methods of meeting expansion, and should take into account the load characteristics of the demand that will be made on the system from year to year.

It would be quite impracticable in the present handbook to follow through such a complex study. However, by adopting certain simplifying assumptions, the problem can be used to demonstrate the effect of certain major factors which are inherent in such economic problems. Among the most important are those arising from:

(a) different patterns of cash flow
(b) widely different asset lives
(c) different degrees and kinds of uncertainty.

In the example which follows the hydro/thermal projects are treated as simple alternatives, representing isolated sources of electric power. No consideration is given to future integration with an interconnected system or to those differences between the sources of power which affect their operational characteristics as part of an expanding system.

Further assumptions are as follows.

1. In both cases the initial and ultimate effective plant installations have the same capacity and the periods of load development between first commissioning and full utilization of the station output are the same.

2. The comparison is confined to a period of load development during the first ten years of operation, and a subsequent period of uniform operation covering the balance of the life assigned to the hydro project.

3. The hydro project has an effective life of 60 years, the electrical and mechanical equipment being replaced after 30 years.

4. The initial conventional thermal project is replaced after 30 years by a nuclear power station of equivalent capacity to which is assigned a life of 30 years.

5. Price levels remain constant during the period under review.

6. Load factors are constant from year to year over the whole period and revenue earned will not change, so that the problem can be evaluated on the basis of the cost either per unit generated or per kW of maximum demand.

For convenience the basic data relating to the alternative projects are summarized in Table 5.6 and subsequent paragraphs.

TABLE 5.6

Type of generation	Hydro	Thermal
Initial plant installation	200 MW	200 MW
Final plant installation	500 MW	500 MW
Period of load growth from 100 MW to 400 MW	10 years	10 years
Total construction expenditure on first commissioning	£50·5 million	£14·1 million
Total construction expenditure when fully developed	£54·9 million	£28·2 million
Output in first year (kWh × 10⁶)	525	525
Output in tenth year and thereafter (kWh × 10⁶)	2100	2100
Running costs in first year of service	£84 000	£1 080 000
Running costs in tenth year of service	£120 000	£4 000 000
Cost of replacements in year 30	£10 000 000	£50 000 000
Running costs after year 30	£120 000	£520 000

The initial construction period is 5 years for the hydro project and 3 years for the thermal project. Construction of the thermal project will begin two years after the hydro project would have begun. The latter date will here be called year 1, so that construction of the thermal project begins in year 3.

The initial plant installation is 200 MW for either project. Additional plant is to be installed in sets of 100 MW coming into operation at the beginning of years 8, 11 and 14.

At the end of year 13 the development phase is completed and the installation for either project is 500 MW. The construction expenditure year by year over this period is shown in Table 5.7.

TABLE 5.7

Year	Cash payments (undiscounted) £ × 10³		
	Hydro	Thermal	Difference (thermal less hydro)
1	5 050	—	− 5 050
2	12 650	—	− 12 650
3	15 100	2 820	− 12 280
4	12 650	8 460	− 4 190
5	5 050	2 820	− 2 230
6	—	—	—
7	1 400	4 700	3 300
8	—	—	—
9	—	—	—
10	1 500	4 700	3 200
11	—	—	—
12	—	—	—
13	1 500	4 700	3 200
	54 900	28 200	− 26 700

The initial maximum demand, in year 6, is 100 MW. This load grows by equal annual increments to 400 MW in year 15. Thereafter the load remains constant. The annual load factor is assumed to remain constant throughout at 60%.

The running costs up to year 15 are shown in Table 5.8.

TABLE 5.8

Year	Cash payments (undiscounted) £ × 10³		
	Hydro	Thermal	Difference (thermal less hydro)
6	84	1080	996
7	84	1404	1320
8	96	1728	1632
9	96	2052	1956
10	96	2376	2280
11	108	2700	2592
12	108	3024	2916
13	108	3348	3240
14	120	3672	3552
15	120	4000	3880

Thereafter running costs are assumed to remain constant to year 30 on a basis of £120 000 per year for hydro plant and £4 000 000 per year for thermal. The difference in favour of hydro in this period is in consequence £3 880 000 per year.

After year 30 it is assumed that the thermal station is taken out of service and its duties taken over by a new nuclear station of equivalent output. The nuclear station in its turn is retired from service at the end of year 60. The hydro alternative might reasonably be expected to remain serviceable for many years thereafter but no account is taken of this as the present value of costs and benefits arising more than 60 years in the future is negligible at any realistic rate of discount.

The capital cost of the nuclear station is £50 million, assumed for simplicity to have been incurred in year 30. Its annual running costs are taken at £520 000.

The operation of the hydro project will entail capital expenditure of £10 million in year 30 on the replacement of electrical and mechanical equipment. The running costs remain unchanged at £120 000 per year. In year 30, therefore, the difference in cash flows for capital outlays in favour of the hydro project is £50 million less £10 million, or £40 million.

The difference in running costs in favour of the hydro project for years 31–60 is constant at £400 000 per year.

Between years 16 and 60 the undiscounted cash flows show a balance in favour of the hydro project of £110 200 000 made up as shown in Table 5.9.

TABLE 5.9

Year		£ × 10⁶
16–30	Difference in running costs at £3 880 000 per year	58·2
30	Difference in capital investment	40·0
31–60	Difference in running costs at £400 000 per year	12·0
Total difference (years 16–60) in favour of hydro project		110·2

61

The undiscounted differential cash flows are shown in Table 5.10. Years 1–15 are obtained by summing the difference columns of Tables 5.7 and 5.8. The figures for the period 16–60 are obtained from Table 5.9.

TABLE 5.10

Year	Undiscounted differences in cash flows (positive figures are in favour of hydro) £ × 10³
1	−5 050
2	−12 650
3	−12 280
4	−4 190
5	−2 230
6	996
7	4 620
8	1 632
9	1 956
10	5 480
11	2 592
12	2 916
13	6 440
14	3 552
15	3 880
Total 1–15	−2 336
Total 16–60	110 200
Difference in favour of hydro project	107 864

The balance shown in favour of the hydro project in Table 5.10 is the amount that would be relevant for a decision if the interest rate were zero. Table 5.11 gives the corresponding discounted figures for discount rates of 4% and 8% per annum. Detailed figures are given for years 1–15.

These figures emphasize the importance of the interest rate in evaluating two projects which involve widely different ratios between capital and running costs. The effect of varying the interest rate becomes particularly important in long-term comparisons. As the rate is increased the present values of benefits and costs decrease faster the more distant they are in time.

TABLE 5.11

Year	Discounted differences in cash flows £ × 10³	
	4% per annum	8% per annum
1	−4 856	−4 676
2	−11 696	−10 845
3	−10 917	−9 748
4	−3 582	−3 080
5	−1 833	−1 518
6	787	628
7	3 511	2 696
8	1 193	882
9	1 374	978
10	3 702	2 538
11	1 684	1 112
12	1 821	1 158
13	3 868	2 368
14	2 051	1 209
15	2 155	1 223
	−10 738	−15 075
16–60	38 440	14 900
Present value at year 1 of difference in favour of hydro project	27 702	−175

Table 5.11 brings out the decisive importance in this instance of the figures used in the latter part of the period under review: the years 16 to 60. Yet it is during this part of the period that the forecasts are likely to prove less reliable.

It should not be overlooked that the uncertainties surrounding the two projects are different. In the case of the hydro project, the main uncertainty is the cost of construction. For the thermal project the greatest uncertainty lies in the future cost of fuel and the security of its supply.

Future technological or economic changes may lead to results which are different from those expected from an analysis based on the best information available at the present day. In this example for instance it has been assumed that in thirty years' time a conventional thermal power station built to current standards will be replaced by a nuclear power station which will take advantage of technological changes over the next thirty years. Bearing in mind that the first nuclear power station in the world only became operational in 1956, any attempt to forecast such changes over the next thirty years is likely to prove hazardous.

Because of these uncertainties it is important to examine the sensitivity of the results of the discounted cash flow analysis to changes in the basic assumptions on which the cost streams are predicted. If, for instance, there are sound reasons for anticipating that fuel prices will rise relatively then the effect of such an increase should be examined.

Again, if the rate of growth in the demand for electricity should be different from that assumed, the comparison between the two projects will be affected. A higher rate of growth will be accompanied by a corresponding increase in the flow of benefits in the form of revenue from sales of electricity. These increased sales can be met in the case of the hydro project by comparatively small increases in investment in future generating plant. In the thermal project, however, the increased investment in generating plant is accompanied by an increase in the annual expenditure on fuel. The net result of such a change is likely to be favourable to the hydro project.

In spite of the difficulties of long range forecasts the effects of future technological and other changes are relevant to a decision taken now and an appraisal of their probable effects may well be critical in reaching a decision. Thus any appraisal should not only incorporate the best forecasts that are available but should also be supported by an examination of the effects of those uncertainties which can be identified.

EXAMPLE 5.6

This example demonstrates an important aspect of economic evaluation, namely, the determination of the most economic period for which an asset such as a piece of machinery should be retained before it is replaced.

The principles which underlie such a decision can be demonstrated by applying them to a simple item of mechanical equipment the economic value of which declines over time. Factors which affect the economic life of such an asset are:

(a) the difference between the original price and the trade-in value; the trade-in value at a particular time is the price expected to be received if the asset is disposed of at that time

(b) running costs, including the cost of maintenance and repairs, insurance etc. during successive intervals in the asset's life, and the down time (the periods during working hours when the equipment is not available for use because of the need for repair)

(c) obsolescence arising from the development of more economical methods of achieving the service the asset provides, or from reductions in the demand for the service; the cost of replacing the service of the asset after various intervals of time is related to the former; a new piece of equipment is likely to have a better performance in terms of output than the old equipment which it replaces

(d) the discount rate.

These principles may be applied to a small excavator. For convenience, the analysis has been limited to six years. The fact that the actual physical life of the excavator could with adequate maintenance be considerably extended beyond this period is not in itself relevant to the determination of its economic life.

It is assumed that the initial cost of the excavator is £4000 and that the required annual working hours are 2000. It is also assumed that wages and fuel will be constant over the asset's life in real terms (that is after adjusting for any expected inflation) and that these costs would be the same for any replacement. For simplicity they are omitted from these calculations. In practice these costs may vary with age or on replacement, and the need to include them in the evaluation cash flows should always be considered.

The trade-in values of the machine after various time intervals are assumed to be as follows.

At the end of year	1	2	3	4	5	6
Percentage of delivery price	75	60	50	40	35	30

It is also assumed that a new machine would be identical to the first machine if bought in years 1 and 2, but that by the end of year 3 a 10% improvement in performance would be expected, and a 20% improvement

65

by the end of year 5. No further improvements are assumed. Trade-in values of replacement machines are expected to have the same percentage values of delivery price as shown above.

The cost of a new machine in real terms is assumed to be as follows at the end of each of the next six years.

Cost (£)   4116   4224   4324   4416   4500   4576

Maintenance, repair and insurance costs in real terms of the machine bought initially and of any replacement are assumed to be as follows over the first six years of life.

Cost (£)   280   360   560   760   960   1080

It will be seen that these costs follow a rising trend and they are indeed an important factor in deciding when a machine has reached the end of its economic life.

Availability as a percentage of the 2000 hours of annual service is assumed to vary as follows, for a new machine and for any replacement, over the first six years of life.

Percentage   95   93   90   86   82   80

With this information it is possible to calculate the cash flows for varying replacement cycles. Table 5.12 shows, in £s, the cash flows for one-year, two-year, three-year and four-year replacement cycles.

The adjustment for increased efficiency assumes that the machine is one unit of a complex so that the surplus or deficit of capacity can be used or made good by adjusting the total number of machines, or alternatively, that the services of a machine can be let or hired at a price reflecting current capital costs; and that maintenance, repairs, insurance, and other costs are not affected. The adjustment then takes the form of a reduction of 10% or 20% in the capital cost and a corresponding reduction in the trade-in value of the same machine.

A six-year period is being assumed. This implies an end to the use of the excavator after six years. Hence there is no renewal in that year

TABLE 5.12

End of year	0	1	2	3	4	5	6
One-year replacement cycle							
Capital cost . . .	4000	4116	4224	4324	4416	4500	
Trade-in value (75% of cost) . . .		−3000	−3087	−3168	−3243	−3312	−3375
Maintenance, insurance and repairs		280	280	280	280	280	280
Adjustment for increased efficiency:							
capital cost . . .				−432	−442	−900	
trade-in value . .					324	331	675
Net cash flow . . .	4000	1396	1417	1004	1335	899	−2420
Two-year replacement cycle							
Capital cost . . .	4000		4224		4416		
Trade-in value (60% of cost) . . .			−2400		−2534		−2650
Maintenance, insurance and repairs		280	360	280	360	280	360
Adjustment for increased efficiency:							
capital cost . . .					−442		
trade-in value . .							265
Net cash flow . . .	4000	280	2184	280	1800	280	−2025
Three-year replacement cycle							
Capital cost . . .	4000			4324			
Trade-in value (50% of cost) . . .				−2000			−2162
Maintenance, insurance and repairs		280	360	560	280	360	560
Adjustment for increased efficiency:							
capital cost . . .				−432			
trade-in value . .							216
Net cash flow . . .	4000	280	360	2452	280	360	−1386
Four-year replacement cycle							
Capital cost . . .	4000				4416		
Trade-in value:							
(40% of cost) . .					−1600		
(60% of cost) . .							−2650
Maintenance, insurance and repairs		280	360	560	760	280	360
Adjustment for increased efficiency:							
capital cost . . .					−442		
trade-in value . .							265
Net cash flow . . .	4000	280	360	560	3134	280	−2025

and when, as is the case with the four-year cycle, the cycle is not complete, the trade-in value at that date is brought into the cash flow. If a longer period of use were under consideration it would be necessary to make assumptions in respect of later years and to extend the cash flow statement.

Assuming a discount rate of 10%, the present values of the cash flows can be calculated to give the following values.

Cycle: one-year   two-year   three-year   four-year
£   7290   6530   6020   6140

These figures do not however allow for the effect of loss of availability due to down time. The adjustment for this should strictly be made to the cash flows by including the additional cost caused by the loss which will depend on the means available to overcome it. An approximate correction can be made by adding to the respective present values the average percentage losses, calculated as in Table 5.13.

TABLE 5.13

Cycle	Percentage loss of time in year: 1	2	3	4	Average
One-year	5				5
Two-year	5	7			6
Three-year	5	7	10		7·3
Four-year	5	7	10	14	9

Adding the average percentages to the present values gives:

Cycle: one-year   two-year   three-year   four-year
£   7650   6920   6460   6690

The result indicates that the three-year cycle is optimal. The present values can be converted into hourly cost equivalents by converting them into annuity equivalents and dividing by the annual hours of use, here 2000. This gives, for the three-year cycle, using the 10% rate of discount, in £s

$$\text{hourly rate} = \frac{6460}{2000}\left[\frac{0\cdot10}{1-\dfrac{1}{(1\cdot10)^6}}\right] = 0\cdot741$$

This can be checked from Table 3 of Appendix F.

68

If it is desired to test the sensitivity of the result to changes in the different cost figures the present value calculations can be performed separately for each component of the cash flows in Table 5.12 and the effect examined of varying the values of the latter.

EXAMPLE 5.7

The kind of problem dealt with in this example is commonly met in the forward planning of an expanding enterprise when the complexity of the situation calls for a trial and error approach to investment decision and a readiness to re-examine long-term plans in the light of new information.

It is the essence of such problems that investment can only be made in discrete steps while the growth in demand for the facility is a continuous process. Thus capacity and demand are continually getting out of step. If it is not possible to bridge the gaps by leasing short term facilities, investment must be kept ahead of demand resulting in a varying amount of unused capacity. Typical examples of this kind of problem are:

(1) selection of a pipeline development programme to supplement and eventually replace road tankers in meeting a growing demand for oil or water

(2) selection of the most appropriate increase in transmission voltage to meet an increasing demand for electricity

(3) selection of the optimum size of plant units for meeting a growing market for concrete aggregates

Such problems involve deciding when each successive investment step should take place and how big the investment in question should be.

The example which follows relates to the provision of covered storage to meet a need which is assumed to be increasing at a constant rate of 1000 cubic metres per year. The alternatives available are:

(a) the purchase of storage buildings as complete units, a characteristic being that the cost per cubic metre of storage falls with increasing unit size

(b) the renting of storage which is assumed to be available at a uniform annual cost of £2 per cubic metre.

A combination of (a) and (b) may be adopted.

The solution of the problem can be considered in two distinct stages. The first concerns the determination of the most economic period for

69

which an asset of specified physical characteristics, in this case a warehouse, should be held before it is replaced. This follows the lines of the previous example, though in this case the time period of the estimated cash flow is extended to infinity. The results of the necessary calculations are summarized in Table 5.14. They refer to a warehouse of 5000 m³ capacity which can be constructed for £40 000 and give the annual equivalent cost on the basis of given assumptions for a range of replacement cycles from 3 to 7 years.

TABLE 5.14

Replacement cycle (years)	3	4	5	6	7
Annual equivalent cost (£)	9129	9082	9000	8897	9063

These figures indicate that the minimum annual equivalent cost occurs if the warehouse is sold after 6 years, and the cycle of purchase and sale then repeated. The annual figures are obtained from the equivalent present values in the way indicated in Example 5.6. They can be interpreted as the minimum level of annual rent for equivalent facilities which the investment must save to be worth while on the basis of the assumptions made.

The second part of the problem deals with the timing of the step-by-step investment and the selection of the size of each investment step. For the reasons stated, a solution to this part of the problem must be sought by trial and error, starting with a scheme which has been chosen intuitively and then proceeding systematically to examine the effect of modifications in the search for greater economy.

In the present example it is assumed that 5000 m³ of storage space are required at the beginning of year 1, and that these requirements will expand steadily at the rate of 1000 m³ per annum until they reach 20 000 m³ at which level they will remain indefinitely. (For simplicity of exposition it is assumed in the calculations that the expansion takes place on the first day of each year.)

The first step is to examine the merits of buying a larger warehouse unit of, say, 10 000 m³ the costs of which are estimated on the basis of cost studies to be seven-fourths as large as those of a 5000 m³ unit. These costs have the same pattern of incidence in time so that the present values and the annuity equivalents vary in the same ratio.

For the optimum replacement cycle of 6 years the average annual cost per cubic metre is:

for a small warehouse (5000 m³) £8897/5000 = £1·779
for a large warehouse (10 000 m³) £15 570/10 000 = £1·557
rental   £2·000

This shows that large warehouses are the least costly providing they are fully utilized. Thus the first tentative scheme might be based upon investment in a series of large units which are kept fully utilized, in combination with the renting of storage space in years when demand exceeds the available capacity. The warehouses built would be replaced every six years. Such a scheme is summarized in Table 5.15 in which capital and operating costs of warehouses owned are expressed as annual equivalents.

TABLE 5.15

Year	Demand for space	Number of large ware-houses	Record of pur-chases and sales (end year dates)	Cost of ware-houses at £15 570 p.a.	Space rented	Cost of rented space at £2 per m³	Total annual cost
	m³			£	m³	£	£
1	5 000	—		—	5 000	10 000	10 000
2	6 000	—		—	6 000	12 000	12 000
3	7 000	—		—	7 000	14 000	14 000
4	8 000	—		—	8 000	16 000	16 000
5	9 000	—	A bought	—	9 000	18 000	18 000
6	10 000	1		15 570	—	—	15 570
7	11 000	1		15 570	1 000	2 000	17 570
8	12 000	1		15 570	2 000	4 000	19 570
9	13 000	1		15 570	3 000	6 000	21 570
10	14 000	1		15 570	4 000	8 000	23 570
11	15 000	1	A sold, B bought	15 570	5 000	10 000]	25 570
12	16 000	1		15 570	6 000	12 000	27 570
13	17 000	1		15 570	7 000	14 000	29 570
14	18 000	1		15 570	8 000	16 000	31 570
15	19 000	1	C bought	15 570	9 000	18 000	33 570
16	20 000	2		31 140	—	—	31 140
17	20 000	2	B sold, D bought	31 140	—	—	31 140
18	20 000	2		31 140	—	—	31 140
19	20 000	2		31 140	—	—	31 140
20	20 000	2		31 140	—	—	31 140

The next step is to consider what changes in the first scheme might be introduced in the hope of securing better economy. Possibilities that might be considered include the following.

(a) Retaining a warehouse for a shorter or longer period than the optimum in order to obtain at an earlier date the benefit of the lower average cost associated with a larger unit. For example, if this scheme included two small warehouses, the one ending its 6 year cycle at the end of year 12, the other at the end of year 13, it can be shown that it would be preferable to shorten the cycle of the latter to 5 years and purchase one large unit at the end of year 12. This would be better than replacing each in turn by a small unit.

(b) Another possibility is to accept some unused warehouse capacity in order to secure the advantage of lower average costs from the larger units. Some schemes for amending the first trial follow.

Scheme 1. Consider the purchase of a small warehouse at the beginning of year 1 for sale at the end of year 5, that is, one year before the end of what would be its optimum life. This would have an annual cost of £9000 (Table 5.14) as compared with £10 000 for rented storage (5000 m^3 at £2 p.a.). The change would show a net benefit of £1000 p.a. and is thus worthwhile.

Scheme 2. Now consider a further step, the retention until the end of year 6 of the small warehouse bought at the beginning of year 1, and the renting of the extra space required in year 6. This would reduce the annual equivalent cost for 5000 m^3 of storage from £9000 to £8897, a saving of £103 p.a. in years 1 to 5.

Under scheme 1, with the purchase of a large warehouse in year 6, the annual cost in year 6 will be £15 570 as in Table 5.15. If, however, scheme 2 is adopted and the additional space is rented in year 6, the corresponding annual cost in that year will be £18 897 (£8897 + 5000 × £2): a net difference of £3327 in favour of scheme 1. The net cost difference over the 6 year period of scheme 2 with respect to scheme 1 can therefore be expressed in present value terms, using an 8% annual discount rate, as

$$£103 \ (1{\cdot}08^{-1} + 1{\cdot}08^{-2} + 1{\cdot}08^{-3} + 1{\cdot}08^{-4} + 1{\cdot}08^{-5}) - £(3327 \times 1{\cdot}08^{-6})$$

This is clearly a negative sum and scheme 2 would be disadvantageous.

Scheme 3. By similar reasoning to that for scheme 1, it can be shown to be worthwhile to buy a small warehouse at the end of year 10 and sell it at the end of year 15.

Scheme 4. Now consider bringing forward the purchase of the first large warehouse from the end of year 5 to the end of year 4. The life of the first small warehouse would be decreased from 5 years to 4 years. This would increase its annual equivalent cost in years 1 to 4 by £9082 − £9000 = £82 (Table 5.14); in year 5 it would reduce costs from £17 000 (9000 + 4000 × 2), as in scheme 1, to £15 570, a net saving of £1430; in subsequent years the annual equivalent cost would not alter—the whole series of replacement cycles would merely be brought forward by one year. The saving from this change would be, in present value terms:

$$£1430 \times 1 \cdot 08^{-5} - £82 \ (1 \cdot 08^{-1} + 1 \cdot 08^{-2} + 1 \cdot 08^{-3} + 1 \cdot 08^{-4})$$

This is clearly a positive sum and this further change would therefore be worthwhile.

Scheme 5. Now consider bringing forward the purchase of the first large warehouse from the end of year 4 to the end of year 3. This would reduce the life cycle of the small warehouse to 3 years. It would increase costs as compared with Scheme 4 in years 1 to 3 by £9129 − £9082 = £47; increase costs in year 4 from £15 082 (9082 + 3000 × 2) to £15 570, a net sum of £488; and leave subsequent costs unchanged. The change would thus be disadvantageous.

Scheme 6. By similar reasoning, it can be shown to be worthwhile to bring forward the purchase of large warehouse C (Table 5.15) from the end of year 15 to the end of year 14 but not to the end of year 13.

This exhausts the changes that seem to merit investigation; the revised plan, which is assumed to be optimal, is summarized in Table 5.16.

In considering an example of this kind it must be remembered that only one decision has to be taken at the present time, that is, the investment to be made in year 1. The calculations have been extended in order to demonstrate the interaction between current and future decisions, recognizing that revised estimates of future demand may change and that it may be necessary to modify a scheme which has been prepared in the light of present day forecasts.

In practice factors to be taken into account when interpreting the foregoing calculations would include the likelihood of continued availability

TABLE 5.16

Year	Demand for space	Large warehouses		Small warehouses		Record of purchases and sales (end year dates)*	Space rented		Total annual cost
	m³	Number	Cost at £15 570	Number	Cost at £9082		m³	cost £	£
0	—	—	—	—	—	a bought	—	—	—
1	5 000	—	—	1	9082		—	—	9 082
2	6 000	—	—	1	9082		1000	2000	11 082
3	7 000	—	—	1	9082		2000	4000	13 082
4	8 000	—	—	1	9082	a sold, A bought	3000	6000	15 082
5	9 000	1	15 570	—	—		—	—	15 570
6	10 000	1	15 570	—	—		—	—	15 570
7	11 000	1	15 570	—	—		1000	2000	17 570
8	12 000	1	15 570	—	—		2000	4000	19 570
9	13 000	1	15 570	—	—		3000	6000	21 570
10	14 000	1	15 570	—	—	A sold, B, b bought	4000	8000	23 570
11	15 000	1	15 570	1	9082		—	—	24 652
12	16 000	1	15 570	1	9082		1000	2000	26 652
13	17 000	1	15 570	1	9082		2000	4000	28 652
14	18 000	1	15 570	1	9082	b sold, C bought	3000	6000	30 652
15	19 000	2	31 140	—	—		—	—	31 140
16	20 000	2	31 140	—	—	B sold, D bought	—	—	31 140
17	20 000	2	31 140	—	—		—	—	31 140
18	20 000	2	31 140	—	—		—	—	31 140
19	20 000	2	31 140	—	—		—	—	31 140
20	20 000	2	31 140	—	—	C sold, E bought	—	—	31 140
...	...	...	...	...	...		...	...	...

* Capital letters refer to large and lower case letters to small warehouses.

of suitable covered storage for rent, confidence in the forecast of future storage needs, the need to accept larger steps in extending purchased storage than in extending rented accommodation etc. Choice would also be affected by the possibility, in the case of purchased storage, of being able to obtain rent for that part of the building not in immediate use, and on the availability of capital for investment.

In a practical case a wider range of alternatives than those investigated above might be considered. For example, a larger warehouse than one of 10 000 m³ might offer economies at a certain point of time. The use of a computer to calculate a range of alternative cost streams by trial and error methods (simulation methods) is obviously worth consideration. Sensitivity tests could be incorporated without difficulty into a program designed for this purpose.

EXAMPLE 5.8
This example provides another illustration of the use of net present value calculations in choosing between alternatives which differ significantly in the relationship between their initial capital cost and subsequent running costs. The example assumes that the project is required to meet a defined rate of growth in demand and the data have been chosen to permit the use of a simple mathematical treatment. As in example 5.5 it will be seen that the choice is sensitive to changes in the discount rate.

The problem is to choose the most advantageous way of meeting a defined growing demand for potable water. Engineering studies offer two alternatives:
  (a) use of an upland source of water entailing construction of a storage reservoir in conjunction with a gravity pipeline
  (b) pumping water from a local river without the need for large-scale storage

Either alternative can be developed in stages and the problem is to select the most economic project and determine its programme of development. Fundamental to the analysis which follows is the assumption that the quality and adequacy of the adopted source is unaffected by the choice of project.

The calculations derive the present value of a stream of expenditures assumed to take place over infinite time. These expenditures include the repeated replacement costs of the assets at the expiration of their assumed lives. As stated the data used have been chosen to illustrate a simple mathematical treatment of present value calculations.

By using a series of discount rates ranging from 4% to 15%, the effect of the cost of capital on the choice is emphasized and it will be seen how the balance of advantage in favour of a particular project changes with the discount rate used.

75

BASIC DATA

Water demand

Two cumecs in year 0 increasing at a compound rate of $4\frac{1}{2}\%$ per annum until year 15 at the end of which growth ceases. The original demand of 2 cumecs will continue after year 0 to be met from existing resources.

Stages of development

Two alternatives are considered: a single stage or two successive stages.

Commissioning dates

Single stage: at the beginning of year 1.
Two stages: at beginning of years 1 and 10.

Capital costs

(a) Gravity supply:

dam and ancillary works	£750 000
pipeline: single stage	£1 600 000
two stage, per stage	£1 200 000

(b) Pumped supply:

intake and pumping station	£400 000
pipeline: single stage	£610 000
two stage, per stage	£440 000
pumps: single stage	£160 000
two stage: stage 1	£130 000
stage 2	£30 000

Assumed lives

Dam, pipelines, pumping station etc.	50 years
Pumps (electrical and mechanical equipment)	25 years

Annual operating costs

(a) Gravity supply
either single or two stage development   £5000

(b) Pumped supply (labour and maintenance)
£15 000 for single stage, or £10 000 for stage 1 rising to £15 000 for stages 1 and 2
Power for pumping   £45 000 per cumec

Table 5.17 summarizes the present values of the expenditures relating to a single stage gravity project while in Table 5.18 similar details are given for a two stage project.

TABLE 5.17. PRESENT VALUES OF EXPENDITURES ON GRAVITY SUPPLY: SINGLE STAGE

Item	Cash amount £ × 10³	Multiplier	Present value at discount rates (£ × 10³)			
			4%	8%	12%	15%
Investment						
dam and ancillary works	750					
pipeline	1600					
	2350	1	2350	2350	2350	2350
Replacement						
dam and ancillary works	750					
pipeline	1600					
	2350	0·1638*	385			
		0·0218*		51		
		0·0035*			8	
		0·0009*				2
Operation						
annual payment	5	25†	125			
		12·5†		63		
		8·3†			42	
		6·7†				33
			2860	2464	2400	2385

* The multiplier in this case is given by the present value of a payment of £1 made at the end of each of t intervals of n years, where t tends towards infinity. It is given by the formula $A = 1/[(1+r)^n - 1]$ where A is the present value, r is the annual rate of interest, and n is the number of years in each interval. The value of $(1+r)^n$ is given by Table 1 of Appendix F. Here $n = 50$.

† For the multiplier see Appendix F, formula for Table 3. For $n \to \infty$ the present value of an annual payment of S at annual rate r becomes S/r. It is assumed in this and the following tables that payments occur on the last day of each year. If desired a single correction may be made to adjust the figures to, for example, a mid-year date, by multiplying the above present value by $(1+r)^{-1/2}$ where r is the annual rate of interest. This is equivalent to placing each annual payment six months earlier.

TABLE 5.18. PRESENT VALUES OF EXPENDITURE ON GRAVITY SUPPLY: TWO STAGE

Item	Cash amount $£ \times 10^3$	Multiplier	Present value at discount rates ($£ \times 10^3$)			
			4%	8%	12%	15%
Investment dam and ancillary works pipeline 1	750 1200		1950	1950	1950	1950
	1950	1				
pipeline 2	1200	0·7026* 0·5002* 0·3606* 0·2843*	843	600	433	341
Replacement dam and ancillary works and pipeline 1	1950	0·1638† 0·0218† 0·0035† 0·0009†	320	42	7	2
pipeline 2	843 600 433 341	0·1638‡ 0·0218‡ 0·0035‡ 0·0009‡	138	13	2	0
Operation annual payment	5	see Table 5.17	125	63	42	33
			3376	2668	2434	2326

* Multiplier is $(1+r)^{-9}$.
† See note * of Table 5.17.
‡ The figures in the cash amount column are the present values of a payment in nine years' time as given by the initial present value of the investment cost of pipeline 2 in the table. The multiplier then applied in each case is given by the formula of note * of Table 5.17.

Table 5.19 summarizes the present values of the expenditure relating to a single stage pumped supply project, while Table 5.20 provides similar information for a two stage project.

TABLE 5.19. PRESENT VALUES OF EXPENDITURE ON PUMPED SUPPLY: SINGLE STAGE

Item	Cash amount £×10³	Multiplier	Present values at discount rates (£×10³)			
			4%	8%	12%	15%
Investment						
intake and pumping station	400					
pipeline	610					
pumps	160					
	1170	1	1170	1170	1170	1170
Replacement						
intake, pumping station and pipeline	1010	0·1638*	165			
		0·0218*		22		
		0·0035*			4	
		0·0009*				1
pumps	160	0·6003*	96			
		0·1710*		27		
		0·0623*			10	
		0·0315*				5
Operation (annual payments)						
pumping	90 in year 0	17·45‡	1570			
		6·76‡		608		
		3·62‡			326	
		2·51‡				226
labour and maintenance	15	†	375	188	125	100
			3376	2015	1635	1502

* See Table 5.17 note *. The value of n for the pumps is 25.
† See Table 5.17 note †.
‡ The present value of power cost of pumping all future demand in excess of 2 cumecs is

$$P\left[\frac{(1+q)-1}{1+r}+\frac{(1+q)^2-1}{(1+r)^2}\cdots+\frac{(1+q)^{15}-1}{(1+r)^{15}}\right]$$

$$+\ P\left[\frac{(1+q)^{15}-1}{(1+r)^{15}}\right]\left[\frac{1}{(1+r)}+\frac{1}{(1+r)^2}+\ldots\right]$$

$$=P\left[\frac{1+q}{r-q}\left(1-\frac{(1+q)^{15}}{(1+r)^{15}}\right)\right]-\frac{P}{r}\left[1-\frac{1}{(1+r)^{15}}\right]$$

$$+\ \frac{P}{r}\left[\frac{(1+q)^{15}-1}{(1+r)^{15}}\right]$$

where P is power cost of pumping 2 cumecs, q is annual rate of demand growth and r is discount rate.
The first line of the above expression represents the period of growth, the second line represents the subsequent period of constant demand.
The example assumes that all water in excess of the 2 cumecs demand in year 0 will continue to be supplied from existing resources and that the cost of the power used in pumping is proportional to the quantity pumped.

TABLE 5.20.   PRESENT VALUES OF EXPENDITURE ON PUMPED SUPPLY: TWO STAGE

Item	Cash amount £×10³	Multiplier	Present values at discount rate (£×10³) 4%	8%	12%	15%
Investment						
intake and pumping station	400					
pipeline 1	440					
pumps 1	130					
	970	1	970	970	970	970
pipeline 2	440	0·7026*	309			
		0·5002*		220		
		0·3606*			158	
		0·2843*				125
pumps 2	30	0·7026*	21			
		0·5002*		15		
		0·3606*			11	
		0·2843*				9
Replacement						
intake, pumping station and pipeline 1	840	0·1638†	138			
		0·0218†		18		
		0·0035†			3	
		0·0009†				1
pumps 1	130	0·6003†	78			
		0·1710†		22		
		0·0623†			8	
		0·0315†				4
pipeline 2	309	0·1638‡	51			
	220	0·0218‡		5		
	158	0·0035‡			1	
	125	0·0009‡				0
pumps 2	21	0·6003‡	13			
	15	0·1710‡		3		
	11	0·0623‡			1	
	9	0·0315‡				0
Operation (annual payments)						
pumping	90 in year 0	see Table 5.19	1570	608	326	226
labour and maintenance	10	7·44§	74			
		6·25§		62		
		5·32§			53	
		4·77§				48
	15	17·56‖	263			
		6·25‖		94		
		2·99‖			45	
		1·90‖				28
			3487	2017	1576	1411

[see p. 81 for footnotes]

The results of the present value calculations are summarized in Table 5.21. The bold figures represent the most favourable project under the selected discount rate. The results are shown graphically in Fig. 5.2.

TABLE 5.21. SUMMARY OF TOTAL PRESENT VALUES OF EX-PENDITURES

Project	Table reference	Present value at discount rates (£ × 10³)			
		4%	8%	12%	15%
Gravity supply					
single stage	5.17	**2860**	2464	2400	2385
two stage	5.18	3376	2668	2434	2326
Pumped supply					
single stage	5.19	3376	**2015**	1635	1502
two stage	5.20	3487	2017	**1576**	**1411**

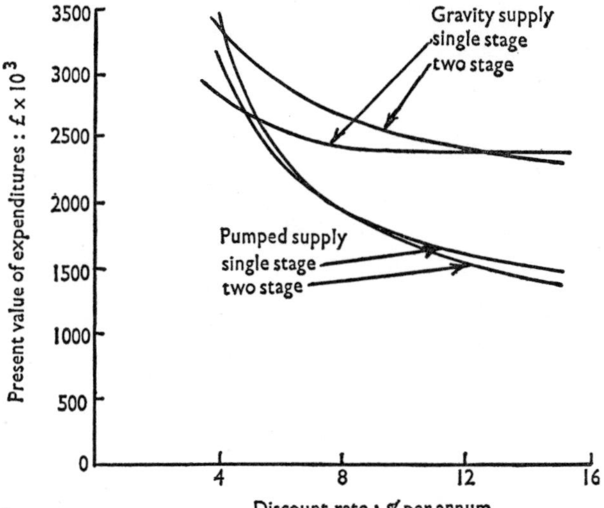

FIG. 5.2

* See Table 5.18, note *.
† See Table 5.17, note *. The value of n for the pumps is 25.
‡ Calculated as in Table 5.18, note ‡.
§ The present value is that of a 9 year annuity (Appendix F, Table 3).
‖ The present value is that of a permanent annuity which begins at the end of 9 years. The present value which such an annuity of £1 per annum would have in 9 years' time is given by the factor $1/r$. This is discounted back to the beginning of year 1 by applying the factor $1/(1+r)^9$.

81

The alternatives considered are characterized in the case of the gravity supply by high capital cost associated with low running costs; in the case of the pumped supply these characteristics are reversed, a smaller investment being followed by high running costs. As the discount rate is raised the advantage turns increasingly in favour of the project in which expenditures occur on average later in time. It will be seen that, on the assumptions made, an advantage in favour of pumping is secured when the discount rate exceeds 5%.

It must be emphasized that the outcome of such an economic assessment is only valid for the specific engineering projects which have been compared. Having reached broad conclusions from such an initial economic study, the engineer would then proceed to examine the economies which might be achieved by modifying the detailed design of the projects.

Chapter 6 Risk and uncertainty

UNCERTAIN DATA

In the examples given in the previous chapter the conclusions have been derived from an analysis of the figures used to represent flows of costs and benefits. The data used and the assumptions made were taken as the best available, and the question of uncertainty in their value was not raised. In practice, however, much of the data used in economic analysis is uncertain in varying degrees, either because of the nature of the evidence itself, or because of difficulties in interpretation.

Project evaluation will usually require the direct physical measurement of certain factors combined with estimates of future events. Some of the direct evidence may not, because of its nature, be precisely measurable. Estimates of future events depend upon an interpretation of what has happened in the past and, since experience shows that the pattern of past events is never exactly repeated, projection into the future can only provide guidance of an imprecise nature.

The limitations of human judgement may also result in imperfect interpretation of, or even failure to recognize, the evidence available. Even in the best conditions different people will attach different weights to particular factors. Consequently two people, both highly trained and competent, may reach different conclusions from a study of the same evidence. Because of the element of personal judgement, in many situations it is not meaningful to speak of a 'correct' or 'true' solution.

PROBABILITY THEORY

Where there are significant uncertainties associated with the result of a project evaluation, it is sometimes useful to list the range of results thought to be likely and to attach to each result an estimate of its probability.

The probability of a result may be expressed in terms of frequency of occurrence, that is, as the proportionate number of times the result would be expected to occur in a large number of trial events.

Probabilities may be estimated from the statistical records of past events. For example, if it is desired to estimate the likely maximum annual flood level of a river as a guide in selecting the height of an embankment, statistical evidence of the type given in Table 6.1 may be used.

TABLE 6.1. RIVER LEVEL RECORDS

Maximum annual flood level above datum	Number of years observed	Probability
36	23	0·46
42	11	0·22
48	8	0·16
54	6	0·12
60	2	0·04
	50	1·00

Similarly, it may be possible to estimate probabilities by conducting experiments using a sampling procedure. For example, the tensile strength of mild steel may be estimated by stressing a series of samples in a tensile testing machine and reading from a dial the load at failure. The testing of several samples will give an indication of the random variations and errors inherent in both the steel and the testing machine. A further check might be performed by calibrating the testing machine to see whether the dial readings were accurate. The experimental results could then be corrected for any errors revealed.

In the social field an excellent example of the successful use of probability theory is given by life assurance. The successful operation of life assurance business depends upon the use of well established tables of proportionate frequency of death at various ages in different populations.

Information about probabilities, of the type in Table 6.1, can be summarized by calculating the mean value as the average of the possible values weighted by their probabilities; and the spread of possible values can be expressed in terms of the standard deviation as shown in Table 6.2.

TABLE 6.2

Recorded annual flood level, h	Probability, p	p × h	(h − H) = d	d²	p × d²
36	0·46	16·56	−6·36	40·45	18·61
42	0·22	9·24	−0·36	0·13	0·03
48	0·16	7·68	5·64	31·81	5·09
54	0·12	6·48	11·64	135·49	16·26
60	0·04	2·40	17·64	311·17	12·45

Mean annual flood level $H = 42\cdot36$	Variance $= 52\cdot44$

Standard deviation* $= \sqrt{(52\cdot44)} = 7\cdot24$

It is often helpful to express probability data in a graphical form. If the predicted result can vary continuously over a particular range it can be illustrated as a continuous curve which relates the various results to their probabilities. Such a curve is shown in Fig. 6.1 which describes the distribution of estimates of the cost of constructing a building in terms of

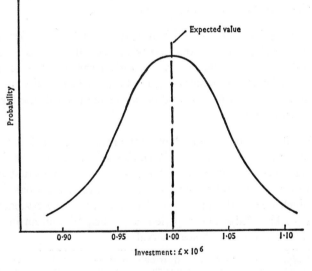

FIG. 6.1

* The standard deviation may be defined as the root mean square value of the deviations or differences from the arithmetic mean.

probabilities.   The mean value here, £1 000 000, is said to be the 'expected' cost.   The greater the difference between a particular value and the mean value, the less probable it is in this case.   For example, this distribution assumes that the cost is more likely to fall between £950 000 and £1 000 000 than between £900 000 and £950 000 and more likely to fall between £1 000 000 and £1 050 000 than between £1 050 000 and £1 100 000.   The probability that a value will fall between any two points is proportional to the area under the curve between those points.

Experience shows that probability distributions are often bell-shaped. Statisticians frequently make use of a symmetrical form, known as the normal or Gaussian distribution, which may be useful in the estimation of uncertain values where the data are believed to be such that this curve offers a good approximation to the true distribution.   This curve is described by the following equation:

$$y = \frac{1}{\sigma\sqrt{(2\pi)}} e^{-(x-\bar{x})^2/2\sigma^2}$$

In this equation the two parameters, $\bar{x}$ and σ represent respectively the mean value and the standard deviation.

In practice, probability distributions may be of many different forms. Some are skew,* as may be the case when the variable has an upper or lower limit.   For example, the level of a river can never be negative but in flood may be much greater than normal.   A probability distribution for the flood level of a river may thus be skewed towards the high values as illustrated in Fig. 6.2.

Very often it is not possible to express the results of a project analysis in the way described above because the statistics necessary for the derivation of a probability distribution are not available.   In such a situation it may still, however, be helpful to present the estimates as a probability distribution which reflects the subjective views formed by the estimator as the result of his knowledge and experience.   A helpful concept in this connexion is that of the confidence interval.

---

* *Skewed distribution.*   Frequency distributions when plotted to rectangular coordinates often present a lack of symmetry.   A distribution which is skewed towards the right, that is positively, is caused by the extremes of the higher values distorting the curve towards the right.   A distribution which is skewed towards the left, that is negatively, is caused by the extremes of the lower values distorting the curve towards the left.

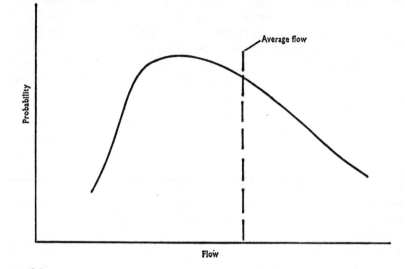

FIG. 6.2

The estimator for the building described in Fig. 6.1 may have expressed 90% confidence that the cost will be between £900 000 and £1 100 000 (that is he considers there is a 9 to 1 chance that the actual cost will be in this range) and 60% confidence that it will be between £950 000 and £1 050 000. If it is assumed that the subjective probability curve has a standard shape (that is that it conforms to the normal or Gaussian curve) it can be derived from this information, using the idea that the probability of a value being within a given range is proportional to the area under the curve. It must be emphasized, however, that such a curve has not the same kind of validity as one that is derived from good statistical data. It should be used only if it is thought to be a helpful way of expressing a personal opinion and the attention of those using it should always be drawn to its subjective origin.

DECISION IN CONDITIONS OF UNCERTAINTY

In appraising a project, the outcome of which is subject to some un- certainty, it may be justifiable to reach a decision (derived by discounting in the normal way) on the basis of the mean values of the relevant quan- tities only, if these values can be obtained. This might be the case if an

7

undertaking were proposing to carry out a number of similar projects and expected that the variations in their results would be distributed randomly and would thus have a high probability of cancelling out. It would also be the case if the range of possible values for the outcome of a project was considered small in relation to either the absolute value of the project or the size of the undertaking, so that the variations expected were not significant.

In other cases, however, the mean or expected value may be an insufficient or misleading basis for decision. The decision-taker may dislike the prospect of a deficit of, say, £100 000 more than he likes the prospect of a surplus of the same amount; similarly he may weigh a cost increase more heavily than a cost saving. For example, the consequences of a serious deficit might involve a small limited company in bankruptcy or at least cause the shareholders to reduce their standard of living as a result of the curtailment of their dividends; for a local authority, a deficit might involve an increase in the rates; and so on. To those affected, these consequences may be more painful than the forgoing of a benefit of equal absolute value. For this reason an undertaking may not be willing to show indifference between two projects which have the same estimated mean value but a different spread of possible values. It may for instance prefer to carry out a project by a method likely to result in a higher mean cost associated with a lower uncertainty rather than use an alternative method which has a lower mean cost but a greater degree of uncertainty; and it may reasonably reject a project which although it shows a high estimated mean surplus of benefit over cost is also subject to a high degree of uncertainty and even the possibility of a deficit.

It is not possible to lay down any general rules to determine whether a project, having a given mean value and a given spread of possible values, should be accepted. Acceptability will depend in part on the attitude to uncertainty of those who take the decision, an attitude which should reflect the views of those whose interests they represent whether they be taxpayers or shareholders. Whatever the decision it should be supported by estimates which indicate the range of possible results which may reasonably be expected.

Another approach to decision-making, where the factors involved in an evaluation are subject to uncertainty, is known as sensitivity analysis.

Initially the appraisal is based on the use of what are judged to be the most likely values; calculations are then made to determine what percentage change in the value assigned to a particular factor would be necessary to reduce the net benefits of the project to zero or by a specified amount. The process is repeated taking all the principal factors in turn and, if thought fit, combinations of factors.

Such an analysis may reveal critical factors in which fairly small errors would seriously affect the economic viability of the project. Once the critical factors have been determined special attention can be devoted to their study.

The two examples which follow illustrate the practical application of the foregoing principles.

EXAMPLE 6.1

An earth dam construction programme allows two summers for the placing of the earth fill. The problem is to decide how much plant should be provided, bearing in mind that such work is sensitive to weather conditions and that it will be desirable to provide additional earthmoving equipment over and above that required to complete the work in an average summer.

Timely completion is important to both the owner of the dam and to the contractor who builds it because on the one hand the owner is concerned to obtain the benefit from his investment when he needs it, while the contractor is anxious to avoid incurring additional costs as the result of delayed completion. In such circumstances it is common for the contract to provide for the payment of a bonus or penalty which can be earned or suffered by the contractor, according to his actual performance. For the sake of simplicity such payments are ignored in the calculations which follow.

In the absence of special benefits, the value to be attached to early completion, as the result of sub-normal rainfall, will be less than the additional costs likely to arise from delay due to abnormal wet weather. In the latter event, unless special steps are taken to meet the situation, the embankment fill will be incomplete when the second summer season ends and a delay of many months will ensue until work can be resumed

89

in the following spring. General considerations thus suggest that the losses arising from delay are likely to outweigh by far the gains from early completion. From this it follows that more earthmoving plant should be provided than is deemed necessary to complete the programme in a season of average rainfall. Calculations are needed to establish the quantities involved.

The problem can be examined systematically in a series of steps, the first being:

(a) to establish a relationship between the output of the plant and monthly rainfall

(b) to establish a relationship between the values of monthly rainfall and the probability of their occurrence.

Guidance on the additional amount of plant to provide may be obtained from a report: *The effect of wet weather on the construction of earthworks.** Here it is suggested that the relationship between plant performance and rainfall can be expressed as

$$L_s = S_s R - K$$

L_s is the plant standing time expressed as a percentage of the plant available time. R is the monthly rainfall in cm. S_s and K are constants for particular types of plant and for different soils. Under British conditions S_s can vary from 0·3 to 6·5 while K may range from 3·1 to 6·5. In the calculations which follow it is assumed that $L_s = 5R - 3·0$.

For convenience of calculation, the example assumes that such a relationship may be applied to the average monthly rainfall over two consecutive summers.

Long term rainfall records make it possible to determine the probability of occurrence of monthly rainfall values. These are summarized in Table 6.3 from which the average rainfall is derived as 10 cm per month.

The direct effects on the works costs incurred by the contractor, as the result of early or late completion, are assumed to be as follows. These effects are expressed in terms of the cost of the earthfill if carried out

---

* Research Report No. 3, 1965, Civil Engineering Research Association.

TABLE 6.3

Monthly rate, cm	Probability of occurrence, %
0– 3	3
3– 6	15
6– 9	20
9–12	30
12–15	25
15–18	5
18–21	2

under average rainfall conditions. The costs are assumed to represent present values.

(1) $A = 0.15\,S$

where A is the benefit from early completion expressed as a percentage of works cost of earthfill and S is the percentage of time saved.

(2) $B = 6 + 0.15\,L$

where B is the loss from late completion expressed as a percentage of the works cost of earthfill and L is the percentage of time lost.

(3) $C = 0.15\,Q$

where C is the cost of additional plant expressed as a percentage of the cost of earthfill and Q is the percentage of plant added.

The next step is to determine the relationship between the time taken for the earthfill and the amount of rainfall. This calculation is set out in Table 6.4 and uses rainfall values to match those in Table 6.3. The results are given for a range of plant quantities and are derived as being proportional to the time ratios in the third column of the table, and inversely proportional to the quantities of plant.

By combining the times given in Table 6.4 over a range of plant quantities with the economic effects of early or late completion it is possible to derive an optimum figure for the amount of plant which should be provided.

The results of these calculations are given in Table 6.5.

From the figures given in Table 6.5 it may be deduced that it would be economically justifiable to provide about 20% more plant than would be necessary for completion in conditions of average rainfall. Table

91

6.5 also shows that while the provision of still more plant would further reduce the risks of delay, the extra costs incurred would be difficult to justify on the basis of the assumptions made.

A calculation on the foregoing lines provides a contractor with useful guidance for cost estimating purposes. However, he would not necessarily feel bound to follow the above conclusion when planning his construction plant programme for the two seasons' work since a decision of this kind would be influenced by his personal attitude to risk. Table 6.5 would, however, be of use to him in selecting some alternative strategy.

For a UK contract he would probably prefer to commence operations with a nominal surplus of earthmoving capacity, adjusting his plant requirements in the light of achievement and keeping open the options of bringing additional plant on to the site or working longer hours in order to make good delays as they arose. (Such options are ignored in the calculations of Table 6.5.) On the other hand many major works of this character are carried out in remote parts of the world where transport time from the UK is important; in such circumstances the contractor must make his assessment of additional plant at the outset and not only of the actual operating plant but also of the supporting facilities, both of men and equipment, to serve that plant.

EXAMPLE 6.2

This example examines the effect on estimated profitability of changes in the basic assumptions used in an economic assessment.

Preliminary estimates suggest that a proposal to install a conveyor belt to reduce dockside handling costs is likely to be economic. What is the probability that this conclusion may prove wrong?

The data needed for appraisal are:

investment cost C	£16 500
savings in handling cost H	£2600 per annum
discount rate r	8% per annum
life of equipment n	10 years
residual value of equipment	nil

TABLE 6.4. RELATIONSHIP BETWEEN RAINFALL AND TIME TAKEN FOR EARTHFILL

Rainfall monthly rate cm	Plant standing time $L_s = 5R - 3{\cdot}0$ %	Time taken as a percentage of time taken if no rain occurs $100/(100 - L_s)$	Time taken as a percentage of time allowed for completion by different quantities of plant		
			P^*	$1{\cdot}2P$	$1{\cdot}4P$
1·5	4·5	105	55	46	39
4·5	19·5	124	66	55	47
7·5	34·5	153	81	67	58
10·5	49·5	198	105	87	75
13·5	64·5	282	149	124	106
16·5	79·5	487	258	215	184
19·5	94·5	1820	965	804	689

* P is the plant required to complete fill in average conditions.

TABLE 6.5. RELATIONSHIP BETWEEN RAINFALL AND NET COSTS

1 Monthly rainfall cm	2 Probability of occurrence %	3 Benefits (− sign) and costs (+ sign) from early or late completion corresponding to different rates of rainfall % of works costs			4 Total expected cost of additional plant plus cost or less benefits* due to late or early completion of work % of works costs		
		P	$1{\cdot}2P$	$1{\cdot}4P$	P	$1{\cdot}2P$	$1{\cdot}4P$
1·5	3	−6·7	−8·1	−9·2	−0·2	−0·2	−0·3
4·5	15	−5·1	−6·8	−8·0	−0·8	−1·0	−1·2
7·5	20	−2·9	−5·0	−6·3	−0·6	−1·0	−1·3
10·5	30	6·7	−2·0	−3·8	2·0	−0·6	−1·1
13·5	25	13·2	9·6	6·9	3·3	2·4	1·7
16·5	5	29·3	23·2	18·6	1·5	1·2	0·9
19·5	2	135·8	111·6	94·4	2·7	2·2	1·9
Plus cost of additional plant						3·0	6·0
Total					7·9	6·0	6·6

* Expected costs or benefits are derived by multiplying figures of column 3 by those of column 2.

Applying these data the surplus of benefit over cost can be obtained by converting the annual savings to their present worth and subtracting the capital cost from the result (see Appendix F, Table 3). That is:

£2600 × 6·7101	£17 446
less capital cost	16 500
net present value of benefit	£946

It will be seen that this sum is small in relation to the investment. The percentage by which each of the different variables has to be altered to reduce the net present value to zero is a measure of the sensitivity of the conclusion.

Simple calculations show that the net present value is reduced to zero if

(a) C increases by £946 or 5·7%
(b) H decreases by £149 or 5·7%
(c) r increases from 0·08 to 0·0925 or by 15·6%
(d) n decreases by 0·8 years or 8%

If C is based on a fixed price quotation it is not likely to alter much. H is less predictable because it is a forecast extending over several years and has to include allowances for work generated and for works reorganization; r can normally only be approximated but its low sensitivity here may make it less likely to affect the decision than changes in the other variables. Although of moderate sensitivity, n is difficult to predict since obsolescence or wear and tear can easily alter the estimated life by two or three years either way.

Such sensitivity tests are clearly helpful but they should be coupled with an appreciation of the probabilities, if necessary subjective, to be attached to different data variations.

One useful approach is to assume that the different values of the variables that are thought worthy of consideration may be combined with one another randomly, and to examine the effect of this. This can be done by attaching an assumed probability distribution to a selected range of values for each of the data and applying a technique termed a 'Monte Carlo' analysis. Table 6.6 shows the probabilities assigned to such a range of values together with the numbers 0 to 99 allocated in proportion to these probabilities.

TABLE 6.6. ALLOCATION OF PROBABILITIES OF NUMBERS 0 TO 99 FOR EACH VARIABLE

Item	Amount	Assumed probability	Allotted numbers
Life, n	9 years 10 years 11 years	0·20 0·60 0·20	0 to 19 20 to 79 80 to 99
Discount rate, r	$7\frac{1}{2}\%$ 8% $8\frac{1}{2}\%$	0·40 0·50 0·10	0 to 39 40 to 89 90 to 99
Savings in handling cost per year, H. .	£2500 £2600 £2700	0·25 0·50 0·25	0 to 24 25 to 74 75 to 99
Investment costs, C	£16 000 £16 500 £17 000	0·30 0·50 0·20	0 to 29 30 to 79 80 to 99

By using random number tables it is possible to select values for each variable corresponding to their probability distributions and to derive a number of net present values which will represent the combined effect of these distributions. For instance, suppose that on randomly selecting a page in random number tables the following first four numbers appear

98   20   01   79

Take the variables in the order listed in Table 6.6: the number 98 is allocated to the life n, giving it a value of 11 years (since 11 years has been allocated the numbers 80 to 99); r becomes $7\frac{1}{2}\%$ because the range 0 to 39 for $7\frac{1}{2}\%$ includes the next number, 20; H becomes £2500, corresponding to the number 01; and C is £16 500 to correspond to the number 79. Using these values the present value of the project can be calculated as in the top line of Table 6.7, at £1789.

Taking successive groups of two digit numbers from the random number tables the procedure is repeated, preferably using a computer program, to give the 100 results set out in Table 6.7. The figures can then be used to compile the frequency distribution shown in Fig. 6.3.

TABLE 6.7
(The present values are found using Table 3 of Appendix F)

n	r	Present value £1 p.a. $\dfrac{1-(1+r)^{-n}}{r}$	H	Present value savings $\dfrac{H(1-(1+r)^{-n})}{r}$	C	Net present value $\dfrac{H(1-(1+r)^{-n})}{r}-C$
year	%		£	£	£	£
11	7½	7·32	2500	18 300	16 500	1800
11	8	7·14	2700	19 300	16 500	2800
10	7½	6·86	2600	17 800	16 000	1800
11	8	7·14	2500	17 800	16 000	1800
11	8	7·14	2600	18 600	17 000	1600
11	8½	6·97	2700	18 800	17 000	1800
11	8	7·14	2700	19 300	16 500	2800
10	8	6·71	2500	16 800	16 500	1300
10	8	6·71	2600	17 400	16 500	900
9	8	6·25	2600	16 200	16 500	−300
10	8	6·71	2600	17 400	16 000	1400
10	8	6·71	2700	18 100	16 000	2100
11	8½	6·97	2600	18 100	16 000	2100
11	7½	7·32	2500	18 300	16 000	2300
10	7½	6·86	2600	17 800	17 000	800
10	8	6·71	2500	16 800	16 000	800
9	7½	6·38	2600	16 600	16 000	600
10	8	6·71	2500	16 800	16 500	300
10	7½	6·86	2500	17 100	16 500	600
10	8	6·71	2700	18 100	16 500	1600
10	8	6·71	2600	17 400	16 500	900
9	8	6·25	2600	16 200	16 500	−300
10	7½	6·86	2500	17 100	16 000	1100
10	8	6·71	2700	18 100	16 000	2100
10	7½	6·86	2600	17 800	16 000	1800
9	8½	6·12	2500	15 300	16 000	−700
11	8	7·14	2600	18 600	16 500	2100
11	8	7·14	2700	19 300	16 000	3300
10	7½	6·86	2500	17 100	16 000	1100
9	8	6·25	2500	15 600	16 000	−400
10	8	6·71	2700	18 100	16 500	600
10	7½	6·86	2600	17 800	16 500	1300
10	7½	6·86	2600	17 800	16 000	1800
9	8½	6·12	2500	15 300	17 000	−1700
10	7½	6·86	2500	17 100	16 500	600
10	7½	6·86	2600	17 800	16 000	1800
10	7½	6·86	2500	17 100	16 500	600
10	7½	6·86	2600	17 800	16 500	1300
10	8	6·71	2500	16 800	17 000	−200
10	7½	6·86	2600	17 800	16 500	1300
10	8	6·71	2600	17 400	16 000	1400
10	7½	6·86	2600	17 800	16 000	1800
10	8	6·71	2500	16 800	17 000	−200
10	8	6·71	2700	18 100	16 000	2100

(continued on facing page)

TABLE 6.7—*continued*

n year	r %	*Present value* £1 p.a. $\frac{1-(1+r)^{-n}}{r}$	H £	*Present value savings* $\frac{H(1-(1+r)^{-n})}{r}$ £	C £	*Net present value* $\frac{H(1-(1+r)^{-n})}{r}-C$ £
9	8½	6·12	2600	15 900	16 500	−600
10	7½	6·86	2600	17 800	17 000	800
10	8	6·71	2600	17 400	16 000	1400
10	8½	6·56	2700	17 700	16 000	1700
10	7½	6·86	2600	17 800	16 000	1800
10	8½	6·56	2600	17 100	16 500	600
10	7½	6·86	2500	17 100	16 500	600
10	7½	6·86	2600	17 800	17 000	800
10	7½	6·86	2600	17 800	16 000	1800
10	8	6·71	2600	17 400	16 400	900
10	7½	6·86	2700	18 500	16 500	2000
11	8	7·14	2700	19 300	16 500	2800
10	8	6·71	2700	18 100	16 500	1600
10	7½	6·86	2600	17 800	16 500	1300
10	7½	6·86	2600	17 800	16 000	1800
9	7½	6·38	2500	15 900	16 500	−600
10	8	6·71	2700	18 100	16 000	2100
10	8	6·71	2600	17 400	17 000	400
10	8	6·71	2500	16 800	16 500	300
10	8	6·71	2700	18 100	16 500	1600
9	8	6·25	2700	16 900	16 500	400
9	8	6·25	2500	15 600	17 000	−1400
11	8½	6·96	2600	18 100	16 000	2100
11	7½	7·32	2500	18 300	16 000	2300
10	7½	6·86	2600	17 800	16 000	1800
11	8	7·14	2600	18 600	16 000	2600
11	7½	7·32	2500	18 300	17 000	1300
10	8½	6·56	2500	16 400	16 500	−100
10	8	6·71	2700	18 100	16 500	1600
10	8	6·71	2700	18 100	17 000	1100
11	7½	7·32	2600	19 000	17 000	2000
10	7½	6·86	2600	17 800	16 000	1800
10	7½	6·86	2500	17 100	16 000	1100
10	8½	6·56	2500	16 400	16 500	−100
9	8	6·25	2500	15 600	16 000	−400
10	8	6·71	2600	17 400	16 000	1400
10	8	6·71	2500	16 800	16 000	800
10	8	6·71	2700	18 100	16 500	1600
9	8	6·25	2600	16 300	17 000	−700
10	8	6·71	2600	17 400	16 500	900
9	8½	6·12	2600	15 900	17 000	−1100
10	8	6·71	2600	17 400	17 000	400
10	7½	6·86	2600	17 800	17 000	800
10	7½	6·86	2500	17 100	16 500	600
11	7½	7·32	2700	19 800	16 000	3800

(*continued overleaf*)

TABLE 6.7—*continued*

n	r	Present value £1 p.a. $\dfrac{1-(1+r)^{-n}}{r}$	H	Present value savings $\dfrac{H(1-(1+r)^{-n})}{r}$	C	Net present value $\dfrac{H(1-(1+r)^{-n})}{r}-C$
year	%		£	£	£	£
9	$7\frac{1}{2}$	6·38	2500	16 000	17 000	−1000
11	8	7·14	2500	17 800	16 500	1300
9	$7\frac{1}{2}$	6·38	2600	16 600	16 500	100
9	$8\frac{1}{2}$	6·12	2500	15 300	17 000	−1700
10	$8\frac{1}{2}$	6·97	2700	18 800	16 500	2300
10	$7\frac{1}{2}$	6·86	2700	18 500	16 500	2000
10	8	6·71	2600	17 400	16 000	1400
10	8	6·71	2600	17 400	16 000	1400
10	8	6·71	2700	18 100	16 500	1600
10	8	6·71	2500	16 800	16 000	800
10	8	6·71	2500	16 800	16 500	300

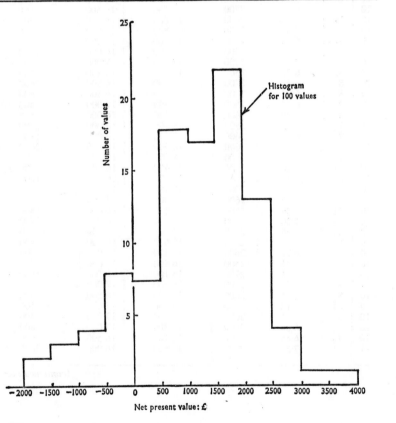

Histogram for 100 values

Net present value: £

FIG. 6.3

This type of calculation can readily be handled by a computer and the example given shows that out of 100 trials, 17 indicate that even though small ranges have been used for the variables, the project as described would be uneconomical. In the example used all the combinations of probabilities assumed could in fact have been covered by fewer calculations but the value of this method will be appreciated in problems where the number of variables and the possible range of each is large.

The method of treatment leads to a better understanding of the uncertainty surrounding the results obtained from the use of average values of data than is possible from simple sensitivity tests. However, a major difficulty in using a 'Monte Carlo' analysis is the determination of the probability distributions which should be applied to the data involved. It is not always possible to construct a frequency distribution curve or histogram from actual observations. More frequently such assessments depend upon judgement, as in the above example, rather than direct measurement.

Chapter 7   Valuations

This chapter contains more examples. The first two demonstrate methods of reducing tenders to a common basis for the purpose of evaluation. The third example shows how discounting methods may be applied in estimating a fair purchase price for a going concern. The fourth example shows how the purchase price of an asset may vary over a wide range, depending upon circumstances and the bargaining position of the purchaser. The last example deals with a method sometimes used for the valuation of assets, the ownership of which is to be transferred under statutory authority.

ADJUDICATION OF TENDERS

The main purpose of tendering procedure is to secure economies in project costs by establishing competition between firms willing to enter into a contract for the performance of specified work. If competition is effective, tenderers will be encouraged to use their resources and expertise efficiently, and thus reduce costs, in the hope of securing the contract and making a profit.

If the objectives of the competitive tendering system are to be realized, the evaluation of tenders must ensure that like is compared with like and that all factors which may affect the performance of the contract are taken into account. To assist evaluation the tender documents, and particularly the technical specifications, should therefore be devised to facilitate a fair comparison between tenders.

Sometimes—and this is especially so when new and rapidly-evolving technology is concerned—economies in cost can be secured by permitting the tenderer to offer his own design proposals which take advantage of his special experience and resources.

In the selection of electrical or mechanical equipment, an important aspect is the performance, expressed in terms of operating efficiency, offered by the tendering firms. The fair evaluation of such tenders throws a special responsibility on the engineer.

Examples 7.1 and 7.2 illustrate two of the more important features of tender adjudication. Example 7.1 compares alternative tenders offering different first cost and performance guarantees. Example 7.2 compares alternative construction programme proposals by examining their effect on the costs of financing a project.

EXAMPLE 7.1

An undertaking receives tenders for the supply and installation of pumping equipment to deliver a constant quantity of water at a constant head. The tenderers quote first costs for the supply and erection of the equipment, and performance guarantees. Which offer is the most advantageous?

The analysis which follows is based on the assumption that equal weight can be attached to the prices and technical guarantees submitted by each of the tendering firms. The basic information is summarized in Table 7.1.

TABLE 7.1

Firm	Tender sum £	Guaranteed efficiency %	Minimum power factor
A	74 150	69·0	0·85
B	75 525	76·6	0·85
C	77 750	69·8	0·85
D	77 920	73·1	0·85
E	78 700	77·3	0·85

In addition to the information provided by the tenders, it is necessary to make assumptions regarding the cost of power, the life of the plant and the discount rate which will normally be the effective cost of the capital employed.

The assumptions are:

electricity tariff (2 part)	£10 per kVA of maximum demand per annum
	£0·004 per kWh of energy consumed
constant output	700 kW (1·7 cumecs at 42 m head)
discount rate	8% per annum
life of plant	25 years

It is now possible to determine which tender offers the lowest sum on the basis of the initial investment cost plus the present value of future power costs.

The power costs can be expressed in terms of the guaranteed efficiency n as follows.

$$\text{kVA cost} = \frac{700 \times £10}{n \times 0·85} = \frac{£8235}{n} \text{ per annum}$$

$$\text{kWh cost} = \frac{700}{n} \times 8760 \times 0·004 = \frac{£24\,528}{n} \text{ per annum}$$

$$\text{total cost} = \frac{£32\,763}{n} \text{ per annum}$$

$$\begin{aligned}\text{present value of sum of power costs*} &= \frac{£32\,763 \times 10·675}{n} \\ &= \frac{£349\,735}{n}\end{aligned}$$

The total present values of the tenders can then be set out as in Table 7.2.

TABLE 7.2

	A	B	C	D	E
Tender sum £ × 10³	74·2	75·5	77·8	77·9	78·7
Power costs £ × 10³	506·9	456·6	501·1	478·4	452·4
Total present value £ × 10³ . .	581·1	532·1	578·9	556·3	531·1
Ranking	5	2	4	3	1

* Using Appendix F, Table 3, for the present value of a 25 year annuity.

The above assessment is illuminating. A, the lowest tender on price, becomes the highest, the second and fifth ranked tenders for price compete closely for first place. Comparison between B and E shows an advantage to E of £1000. This small amount is not significant since it could easily be eliminated by errors in predictions of electricity tariffs, discount rate, maintenance costs, or the effectiveness of the pumps in meeting their guaranteed efficiencies. Moreover, the basis assumed leads to the conclusion that the highest tender A is only some 9% above the lowest tender. Such a close ratio emphasizes the importance not only of examining the effect of introducing small changes into the basic assumptions on which the analysis rests (such as the discount rate) but of taking account of some of the imponderable factors which do not appear in a simple analysis of tenders. Such factors would include the known performance, experience, and financial stability of the firms concerned. These are matters on which it may well be proper for the engineer to comment in his report when making a recommendation for the award of the contract.

A final point concerns the information supplied to tenderers. There are advantages, particularly should it later be necessary to make a recommendation against acceptance of the lowest tender, in advising firms of the basis on which their tenders will be assessed. Thus the formula used in the foregoing analysis could be included in the tender documents, i.e.

$$C + \frac{349\ 735}{n}, \text{ or say, } C + \frac{350\ 000}{n}$$

where C is the tender sum and n is the guaranteed efficiency.

Tenderers would then be able to prepare their own optimum designs to minimize this cost and more efficient as well as more equitable tendering would be achieved.

Inevitably, the determination of the most favourable tender depends upon acceptance at their face values of the efficiencies guaranteed by the tendering firms. Such figures, however, represent a promise only of what the firms hope to achieve and since the actual efficiency as determined subsequently from site tests may differ somewhat from the figure used in the selected tender, the cash value of this difference, whether higher or lower, should be taken into account in arriving at the sum eventually paid to the contractor.

Since such an adjustment can only be made after the installation is completed, the method of calculating the bonus or penalty should form an essential part of the contract.

EXAMPLE 7.2

Three tenders for the construction of navigation locks are based on the employment of different programmes and methods of construction. How can the differences in the economy of these proposals be assessed?

The three proposals include statements of the estimated cash expenditures which would fall due for payment at the end of each quarter during the construction period, as shown in Table 7.3.

TABLE 7.3

Proposal		A $£ \times 10^3$	B $£ \times 10^3$	C $£ \times 10^3$
Quarter	1	340	340	260
	2	160	200	160
	3	120	180	100
	4	200	220	130
	5	350	230	160
	6	400	200	340
	7	100	170	400
	8	—	100	100
		1670	1640	1650

The quarterly cash expenditures as listed above reflect the different methods which the tenderers propose to employ. These are summarized below.

Proposal A can be completed in 21 months by using a heavy concentration of earth-moving plant and by utilizing off-site concrete production facilities.

Proposal B is based on obtaining concrete aggregate by crushing rock from the excavation and on setting up an on-site concrete production plant. The programme requires heavy capital expenditure during the early stage of the contract before concreting can begin. The estimated time for completion is 24 months.

Proposal C is similar to Proposal A but, because it is based upon a more leisurely excavation programme, it will take 24 months.

The analysis which follows assumes that the engineer has made a careful check on the different construction proposals and costs and is satisfied that they are mutually consistent. It further assumes that if the construction period is shortened the project will commence to earn revenue at an earlier date,* and that the additional revenue in quarter 8 to be secured by the adoption of proposal A is estimated at £50 000.

Construction is to be financed by means of an international bank loan. The terms of this loan include a commitment charge at the rate of $\frac{1}{2}\%$ per year, payable on the total value of the loan regardless of the amount actually utilized.† The interest payable on the amount drawn is 6% per year. Both charges are to be added to the loan during the first two years of the loan's currency. Repayment of the loan, plus interest, commences at the end of the two-year period. Under the terms of the loan it cannot be used for any other project, and the granting of the loan does not reduce the availability of finance for other projects. The relevant interest cost in comparing the proposals is therefore the effective rate charged on the loan.

Implicit in the analysis which follows is the assumption that both the costs and the completion dates associated with the three proposals are reliable, as is the estimate of additional revenue to be obtained from proposal A. The principal uncertainty surrounding such an appraisal is likely to be the accuracy of the expenditure forecasts arising from work which may prove to be more difficult that anticipated.

---

* For purposes of illustration it has been assumed that the works will have an indefinitely long life, and the earlier revenue benefits to be derived from proposal A will have no effect on benefit earned later, which will be the same for each of the three proposals.
† It is usual for the lender to make a commitment charge to compensate for the cost of holding funds at the disposal of the borrower. This charge may be levied on the whole of the loan, or as in the case of the World Bank, on the undrawn portion only.

The most straightforward approach in comparing the proposals is to determine in each case the total loan debt at the end of the two-year period. The calculation which follows in Table 7.4 translates the quarterly expenditures to this date by adding accumulated interest (which is assumed to be calculated quarterly). The commitment charges are added, though in this case they do not affect the result, being the same for each proposal. Inclusion of the revenue benefit from early completion gives an estimate of total debt at the end of the initial two-year period.

TABLE 7.4. TOTAL INITIAL COST OF ALTERNATIVE PROPOSALS

Proposal	A $£ \times 10^3$	B $£ \times 10^3$	C $£ \times 10^3$
Accumulated value of quarterly expenditure at end of two-year period			
Quarter 1	376	376	288
2	175	218	175
3	129	194	108
4	212	233	138
5	366	240	167
6	412	206	350
7	101	173	406
8	—	100	100
	1771	1740	1732
Commitment charge	18	18	18
Saving from early completion .	−50		
Accumulated value of net outlays at end of eighth quarter . .	1739	1758	1750

From the foregoing analysis it is seen that the additional capital expenditure involved by reducing the construction period from 24 to 21 months under proposal A is more than compensated by the additional revenue received.*

---

* It should be noted that if the benefit from receiving revenue early under proposal A had arisen in an earlier period (say in quarter 5 and thereafter) it would have been necessary to estimate the interest rate to be used in accumulating it to the end of quarter 8. This might be above 6%, the rate payable on the bank loan, since the revenue could be used (unless the loan agreement required it to be applied to the reduction of the loan) for other financing on which the return might be higher.

The calculation summarized in Table 7.4 represents a simplification of the more elaborate method of calculation used in Chapter 8 to demonstrate the effect of interest over a five-year construction period. The choice of calculation is a matter of judgement. Where long construction periods are involved, approximations of commitment charges as made above may not be found sufficiently accurate and the treatment used in Chapter 8 would in that case be more appropriate.

ECONOMIC VALUE OF ASSETS

The value of assets can be represented by the present value of the cash flows expected to be derived from them. When the purchase price of an asset has to be determined it is necessary to consider the amount which would have to be paid to compensate the owner for its loss. This can be expressed as the difference between the present value to the owner of two cost benefit streams, the one representing the best course of action if the assets are retained, the other representing the best course of action if the assets are disposed of.

The need for such estimates arises when it becomes necessary to make a valuation of commercial undertakings, for example to determine the maximum offer price that a purchaser should make. The following example provides a simple illustration.

EXAMPLE 7.3

What is the value of a pulp mill to a prospective purchaser? The estimated data for valuation include:

average cash amount that can be withdrawn from mill, net of tax	£800 000 per annum
net-of-tax return alternatively available to purchaser	6% per annum
working capital* required for operating the mill	£1 500 000
expected life of mill	10 years
residual value (mostly land) at end of life	£1 000 000

---

* Working capital represents the finance needed to bridge the time gap between purchases of incoming materials and labour and the revenues from sale of the outgoing products, and to maintain a minimum cash balance for contingencies.

107

An estimated value can be obtained from the following present value calculation.

The present value of the net receipts are:

profits £800 000 × 7·360*	£5 888 000
residual value (land and working capital)	
£2 500 000 × 0·558†	£1 395 000
	£7 283 000
Deduct initial value of working capital	£1 500 000
Value of mill	£5 783 000

Although this method of calculating the value of a going concern appears to have mathematical precision, and is based on generally accepted principles for valuation, it is clear that the validity of the estimated value will depend upon the experience and judgement shown by the valuer in selecting his data, and upon the degree of confidence which can be attached to them. It should be noted that neither the original cost of the mill nor its replacement cost have any direct bearing on the valuation. However, the cost of an alternative site plus the cost of building a new mill fixes an upper limit to the value after allowance is made for loss of revenue during the building period.

EXAMPLE 7.4

What is the residual value, at the end of a contract, of a pontoon bridge if the contractor has no further use for it but is required under his contract to remove the bridge on completion?

The bridge cost £35 000 to supply and £5000 to erect. It is in good order and 3 years out of an estimated 12 year life have elapsed. The cost of supply and erection has relevance only in giving some indication of what a similar bridge would cost now. The relevance of the other facts will depend on the circumstances under which the bridge may be disposed of. The following possibilities may be considered.

$* \dfrac{1-(1+r)^{-n}}{r}$ where $r = 0·06$ and $n = 10$ (see Appendix F, Table 3).

$† (1+r)^{-n}$ where $r = 0·06$ and $n = 10$ (see Appendix F, Table 2).

(1) If the only perceptible use is sale as scrap for £3000, and if dismantling and transport will cost £5000, the value to the owner is negative since it represents a charge against the construction project of £2000.

(2) A prospective buyer A, who has a need for the same type of bridge, might be prepared to pay some figure up to the present supply cost of a new bridge, say £40 000, less the cost of dismantling and transport, say £5000. In this case a bargain may be struck at any price between £35 000 (the equivalent value of a new bridge), and −£2000 (the loss to the owner if he sold it for scrap).

(3) If A learns that another prospective buyer, B, is willing to offer £28 000, he may then be prepared to bid, say, £30 000 to secure the bridge. Competition to buy between A and B may drive the price up to £35 000.

The figures in (2) and (3) assume that the bridge is bought in situ, A or B paying the cost of dismantling and transport.

STATUTORY VALUE OF ASSETS

It is not always the case that assets are valued on the basis of their expected future benefits. In accounting practice, for example, statutory rules have to be applied to the valuation of assets in connexion with taxation computations. For the purpose of financial reports prepared to satisfy the rules of company law it is normal to value fixed assets on the basis of their first cost, less an estimated allowance for depreciation.

Special rules may be laid down by statute for the valuation of an undertaking as a whole, as in various nineteenth century Acts of Parliament that authorized the public purchase, under specified circumstances, of private utility undertakings; and as in the case of the Acts that created nationalized industries.

When presented with a valuation problem under statutory rules, the engineer's first duty is to find out whether any terms have been laid down governing the method of valuation. If so, and provided the terms are sufficiently precise and detailed, it becomes a straightforward matter of interpretation, as in the electricity supply example which follows. Where no valuation terms are specified, it is sometimes possible for the two parties to agree on some basis, for example one defined in some statutory document. No valuation should be attempted without some such prior agreement on the basis to be followed.

One of the first examples in the United Kingdom was that embodied in the Tramways Act of 1870. Concessions to operate tramways were granted for specified periods, at the end of which the undertaking could be taken over by a public body (for example, a municipality) at the 'then value' of the physical assets. Any payment for goodwill was expressly excluded and this basis was often called 'tramway terms'.

The Tramways Act did not provide a definition of the expression 'then value' but subsequent practice has been to relate it to the initial cost, the unexpired life, and the total life. If depreciation is on the straight-line basis (that is, the value of fixed assets is assumed to fall linearly, year by year) the 'then value' would be the initial cost multiplied by the ratio which the unexpired life bears to the total life. Thus if an electric generator with an estimated useful life of 30 years, and no residual value, was installed ten years ago at a cost of £30 000, its present value would be estimated at £20 000.

If such a basis is specified (or can be agreed between the parties concerned in the transaction) the procedure is to obtain a complete schedule of the assets with their installed costs (including capitalized interest during construction, that is interest paid and added to the balance sheet value of the asset at the end of the construction period), and their dates of installation or commissioning. The useful life of each class of asset, if this is not specified, should also be agreed between the two parties. The computation is then a purely arithmetical one, and can be seen from the simplified example which follows.

EXAMPLE 7.5

Calculate the value of a steam power station on the above basis.

The data for such a calculation will be known from the financial records and the specified component lives and rates of depreciation. These data and the valuation based on them are shown in Table 7.5.

TABLE 7.5. TRANSFER OF STEAM POWER STATION IN 1968

	A	B	C	D	E
	Cost of installation £ × 10³	Date of installation	Useful life years	Unexpired life years	Valuation A × D/C £ × 10³
1. Freehold land .	20	1951	∞	∞	20
2. Leasehold land, buildings, sidings, etc.	80	1951	40	23	46
3. Boilers . . .	150	1952	20	4	30
4. Turbo-alternator set No. 1 . . .	100	1952	25	9	36
5. Turbo-alternator set No. 2 . . .	100	1961	25	18	72
6. Transformers and switchgear . .	50	1953	25	10	20
Total	500				224

A valuation on the above lines based purely on book-keeping informa-tion, while possessing the merit of simplicity, suffers from serious imperfections, and would be regarded by an economist as unsound in theory. In the first place, such calculations take no account of the economic value of the assets based on their expected earnings (which at worst is their net selling value); for this reason the depreciation of the assets does not fairly indicate the loss to their owner.

Secondly, the valuation takes no account of the value of the organization as a whole. A well-functioning organization may have a value based on expected earnings that is substantially higher than the values of the individual assets that can be identified separately. This excess value is sometimes called goodwill. If it is ignored the owner is again deprived of value without compensation.

Thirdly, the valuation is based on original (historic) costs, whereas the replacement cost of similar plant to provide a similar service may be higher, for example, as the result of inflation, or lower, due possibly to technological progress. If the transfer terms specified that the valuation should be based on the present-day cost of replacement by identical plant (with appropriate allowance for expired life) it might be necessary

111

to approach the original manufacturers for an estimate of present costs. It could well happen that plant of the type installed was no longer in manufacture. Recourse would then be had to official price indices (such as those compiled by the Board of Trade) appropriate to the class of plant under consideration. This would allow for inflation but not for technological progress.

In general, a valuation needs to be backed by a physical inspection. For each major asset or group of assets to be taken over, the inspection should ensure that the asset is in existence and functioning as listed, that it corresponds to its description, and that its condition is reasonable in relation to the life expiry assumed in the valuation.

VALUATION OF INDUSTRIES FOR NATIONALIZATION

When a large utility such as electricity, gas or transport is nationalized, the terms of the valuation or 'consideration' are usually laid down in the Act and subsequent enactments. In the case of the 1947 nationalization of electricity and gas in the United Kingdom, the transfer was chiefly from municipal or company-owned undertakings. Municipal undertakings were taken over on payment by the nationalized industry of the net debts outstanding. The company valuations were based on stock exchange quotations, on specified dates, of the companies' shares, or, if the shares were in fact not quoted on the stock exchange, on an estimate of what the quotation might have been.

A more recent United Kingdom nationalization was that of the steel industry in 1967. The method of valuation employed was that of share price quotations as in the case of the company-owned electricity and gas undertakings mentioned above.

The latter method of valuation has the advantage of saving the enormous amount of effort and skill that would otherwise be involved in making separate valuations and has the theoretical merit that share values represent independent estimates of expected future earnings.

The market price can be assumed to measure the value per share that an owner sets on a marginal change in his holding (since otherwise he would buy or sell some shares). It can, however, be objected that the market price understates the sacrifice arising from the total loss of his interest.

112

Chapter 8   Budgetary problems and financial planning

An economic evaluation is carried out to establish whether a project is justifiable on economic grounds, or to select from alternative ways of implementing a project that which is the best on economic grounds. The first step is to carry out, in the earlier stages of project planning, a general evaluation on the lines indicated in preceding chapters. On the basis of this, and of engineering and financial judgement, a provisional decision is made. Further investigation will follow, leading to a more detailed evaluation. If the more detailed studies justify the results indicated by the earlier evaluation, a firm decision may be made to proceed with the project. If on the other hand this is not the case, the original assumptions will have to be reconsidered and an alternative approach tried, based on the original or on a fresh evaluation.

In the earlier stages of evaluation it would not be usual (or practicable) to work out the financial plan for each possible project in great detail. When, however, a particular project has been provisionally selected, it is necessary to examine in more detail the exact form that the financing will take and the nature of the choices that are available with respect to this. At this stage it is possible, as stated above, that difficulties may appear that were not foreseen in the original evaluation, making it necessary to review the provisional decision. Eventually, when a firm decision to proceed with a project has been taken, a detailed financial plan must be prepared, of which the cash flows of the project form a basic element. This plan will take into account the flows of cash from lenders, if money is to be borrowed; from shareholders, if shares are to be issued; and from revenue earned from the project (unless it is of a type that does not yield cash revenues). Similarly the financial plan must provide for the payment of interest or dividends on the finance raised, and for repayment of money

113

borrowed. For this purpose detailed chronological schedules, or budgets of receipts and payments, must be drawn up.

This financial planning and budgeting demonstrates, as stated in Chapter 1, the detailed financial implications of judgements made in the economic evaluation.

As stated in Chapter 2, when finance has to be rationed the discount rate used in the economic evaluation may be higher than the external cost of finance. The detailed financial planning must, however, allow for the payment of the actual interest due under loan contracts. Earnings in excess of such payments will appear in the financial plan in the form of surpluses that can be withdrawn from the project by the owners.

LOAN FINANCE

A large part of the finance needed to execute large projects is likely to be raised by borrowing. Where this is the case, part of the financial planning will be concerned with the determination of the amounts of finance that are to be raised in this way, and with the timing of the borrowing.

The following example illustrates the kind of detailed planning that might precede an application (for example, to the Government) for loan finance for a project.

EXAMPLE 8.1

For illustrative purposes the hydroelectric project described in Example 5.5 is used. It is assumed that a loan is to be raised to cover 75% of the total initial investment, the balance being met from the internal resources of the authority responsible.

The loan terms include a commitment charge at the rate of 0·5% per annum compounded at half-yearly intervals on the money which has been reserved for the loan, together with an interest charge of 5·5% per annum payable half-yearly on all outstanding debt. Capital repayments, and payment of interest and the commitment charge, are to be deferred until year 6, after the completion of construction. Inflation of costs at 3% per annum is to be allowed for. What will be the total debt at the end of year 5?

114

In Example 5.5 it was assumed for the purpose of evaluation that there would be no inflation. This is not inconsistent with the assumption now made. The aim in Example 5.5 was to establish the merits of alternative projects at varying possible rates of interest. These rates of interest can be defined as real rates, that is as net of inflation. The calculations then show which is the more economical project at a given real cost of capital. The rate on which the decision is based will take into account the method by which the finance is expected to be raised, and the effect of this on the overall rate of return that should be calculated as indicated in Chapter 2.

In the example now to be considered the hydroelectric project has already been chosen. It is now necessary to establish the actual pattern of expected receipts and payments, including amounts borrowed and repaid, and interest payments on these, in order to settle the loan contract and draw up detailed budgets. The interest rates that are now involved are contractual rates, actually to be paid.

The loan needed to finance 75% of the total initial investment includes elements to cover the following:
  (a) estimated inflation of costs
  (b) commitment charge on the total amount of the loan
  (c) interest on the portion of the loan that has been drawn
  (d) further interest charges from the capitalization of (b) and (c)—that is, the addition of (b) and (c) to the loan total during the period when repayment is deferred and before interest payments begin.

Basing estimates on 75% of the cash outlays on construction expenditure shown in Table 5.6, and assuming that annual expenditures occur at each mid-year point, the annual cash outlay, allowing for inflation, will be as follows.

Year		Amount £ $\times 10^3$
1	$5\,050 \times 0{\cdot}75 \times 1{\cdot}03^{0{\cdot}5}$	3 845
2	$12\,650 \times 0{\cdot}75 \times 1{\cdot}03^{1{\cdot}5}$	9 917
3	$15\,100 \times 0{\cdot}75 \times 1{\cdot}03^{2{\cdot}5}$	12 194
4	$12\,650 \times 0{\cdot}75 \times 1{\cdot}03^{3{\cdot}5}$	10 522
5	$5\,050 \times 0{\cdot}75 \times 1{\cdot}03^{4{\cdot}5}$	4 327
Total net cash expenditure		40 805

115

The assumption of mid-year instead of end-year dates (as in Example 5.5) is another example of the greater precision in points of detail at this stage.

The total amount of the loan can now be calculated by assuming that the whole of the loan bears the commitment charge throughout the five year construction period at the rate of 0·25% per half year, and that superimposed on this charge is the interest on the amount drawn to meet construction expenditure, 5·5% per annum, or 2·75% per half year.*

The calculation is as follows.

Commitment charge
$$L \times (1 \cdot 0025^{10} - 1) = 0 \cdot 0253 \, L$$
Construction expenditure plus
superimposed interest £ $\times 10^3$

$3\,845 \times 1 \cdot 0275^9$	4 908
$9\,917 \times 1 \cdot 0275^7$	11 991
$12\,194 \times 1 \cdot 0275^5$	13 966
$10\,522 \times 1 \cdot 0275^3$	11 414
$4\,327 \times 1 \cdot 0275$	4 446
	——
	46 725

whence $0 \cdot 9747 \, L = 46\,725$ or $L = 47\,938$

Thus the total loan required, £47 938 000, represents an increase of about 17½% on the sum of the cash amounts, £40 805 000, actually expended on construction.

It should be noted that if the annual construction expenditure were regarded as spread evenly throughout each year and was subject to the continuous compounding of interest, the amount of the loan would be

---

* 5·5% per annum compounded half-yearly can equally well be described as 2·75% per half-year compounded half-yearly. It is convenient to use the latter form, since the standard tables can then be used with *n* equal to the number of half-years (see Appendix F).

slightly higher than that derived from the above calculation. The difference however would not be important and in practice may be ignored.*

REPAYMENT OF LOANS

When finance is raised for a new investment project, it is necessary to consider the timing and the amounts of the required repayments and the availability of resources to meet them. In some cases the operation of the project itself may be expected to generate cash receipts sufficient to repay the loan and it may be decided (or may be legally necessary) to use them for this purpose; under these conditions no special arrangements are necessary to meet the obligations arising from the loan.

In other cases the finance may be more or less permanent without definite provision for repayment. This is often the case where finance is raised by limited companies and it also applies to government loans issued in the form of stock carrying a 'one-way option', that is stock which is repayable only if the government so wishes. In such circumstances again no special repayment plans are required.

Usually, however, loans (as distinct from share capital) are repayable either in a lump sum at a fixed future date or over a number of years (for example, in equal annual instalments). In such cases it is necessary to make careful plans to ensure that cash is available for repayment when required. If repayment has to be made in a lump sum, it may be decided at the appropriate time to raise a new loan of equivalent size, and to use the proceeds of this new loan for repayment of the old; in this case the situation has some features of that in which the original loan was a permanent one, though it should be noted that interest levels vary and the new loan may only be obtainable at the cost of paying a higher interest rate. A company may repay a loan by issuing additional share capital. If the raising of a new loan in the future at an acceptable rate of interest is not

---

* If r is the annual rate of interest per unit, compounded annually, the continuous rate (the annual rate compounded continuously) is defined as δ where $e^{\delta} = 1 + r$. It is common for interest rates to be compounded at annual intervals. If different periods apply they should be stated because if £1 is invested at 10% per annum the annual payment of interest gives £1·100 at the end of the year, whereas two half-yearly payments at 5% per annum give £1·1025. The half yearly interest equivalent to 10% per annum is 4·88%.

to be relied upon, it will be necessary to plan the investment of cash receipts to provide for repayment at the appropriate time.

If a loan has to be repaid in equal annual instalments and the cash receipts expected from the project are not also received in equal annual instalments, it may be necessary to meet some of the repayments when due by use of short-term borrowing or some other financing device.

In the case of local authorities, planning of capital repayment is subject to formal control.   Not only has government approval to be sought for the raising of finance, but certain conditions have to be met concerning repayment.   Local authority investments commonly yield the major part of their return as non-cash benefits in the form of social services (as in the case of roads and schools) so that cash for the repayment of loans has to be provided by further borrowing or by raising money from local taxation, that is from the rates.

When loans are repaid over a period of years there are two main ways of spreading repayments over time:

(1) annual repayment of equal amounts of capital together with interest on the outstanding balance; this is often referred to as straight line redemption
(2) equal annual payments over the whole period of the loan to cover both capital and interest.

The second method is the more common of the two and is generally described as the annuity method.   It is this method which is usually adopted by the World Bank for the funding of its loans (see Appendix D).

The mathematical basis for calculating annual repayments by each of these methods is given in Appendix F.   The practical difference between the methods can best be illustrated by means of a simple example.

EXAMPLE 8.2

Suppose that a loan of £5000 is to be repaid over a period of 5 years at an interest rate of 6%.   Under the two methods the annual payments of interest and capital would be as shown in Tables 8.1 and 8.2.

118

TABLE 8.1.   METHOD 1: EQUAL INSTALMENTS OF CAPITAL

Year	A Opening balance £	B Repayment at end of year £	C Closing balance £	D Interest 6% on opening balance £	E Total payment (B+D) £
1	5000	1000	4000	300	1300
2	4000	1000	3000	240	1240
3	3000	1000	2000	180	1180
4	2000	1000	1000	120	1120
5	1000	1000		60	1060

Total payments   5900

TABLE 8.2.   METHOD 2: EQUAL INSTALMENTS OF CAPITAL AND INTEREST

Year	A Opening balance £	B Repayment at end of year £	C Closing balance £	D Interest 6% on opening balance £	E Total payment (B+D) £
1	5000	887	4113	300	1187
2	4113	940	3173	247	1187
3	3173	997	2176	190	1187
4	2176	1056	1120	131	1187
5	1120	1120		67	1187

Total payments   5935

The total annual payments decline steadily under method 1, whereas they remain constant in the case of method 2 so that the repayments are made later on average.   On the other hand, under method 1, the total repayments amount to £5900 whereas under method 2 they amount to £5935.   (The figures and assumptions should be compared with those in Example 2.1.   It will be seen that these methods are really applications to special cases of the type of calculation set out in that example.)

If a lump sum has to be found to repay a loan at a certain future date, a sinking fund may be established for the purpose and this may be obligatory in some cases.   This involves investing a sum of money each year (usually a constant sum) in an investment which is easily realizable

9   119

for a predictable amount so that the accumulated investment together with interest on that investment that has been re-invested will, at the required date, amount to the sum needed to repay the loan. The method of calculating the annual instalments of a sinking fund is explained in Appendix F.

EXAMPLE 8.3

Suppose that a loan of £5000 at 6% is to be repaid in one sum at the end of 5 years and that a sinking fund is to be established to provide for repayment, the sinking fund investments earning 3% per annum. The calculation is given in Table 8.3.

TABLE 8.3. SINKING FUND INVESTMENT

Year	*A* Opening balance £	*B* Interest earned £	*C* Annual instalment £	*D* Closing balance £	*E* Loan interest paid £	*F* Total payment (*C*+*E*) £
1			942	942	300	1242
2	942	28	942	1912	300	1242
3	1912	57	942	2911	300	1242
4	2911	87	942	3940	300	1242
5	3940	118	942	5000	300	1242

Total payments 6210

Thus the borrower has to pay £1242 per year to meet the interest on the loan and to provide for the repayment of capital.

It will be apparent that unless the sinking fund investments earn at least as much as the interest cost of the loan itself, the use of the sinking fund method will increase the effective cost of the loan. In the above example the direct cost of the loan, 6% per annum, is increased to an effective 7% per annum, the value of r in the equation:

$$5000 = \frac{1242}{(1+r)} + \frac{1242}{(1+r)^2} + \cdots + \frac{1242}{(1+r)^5}$$

If the sinking fund earns the same rate of interest as that paid on the loan, the annual service of the loan will be the same as with the annuity method which is, in effect, a sinking fund method that operates by buying back the loan that is to be repaid.

The overall effect of loan financing on the time spread of expenditure can be expressed in terms of its impact on the cash flow of a project.   Example 8.4 illustrates how the outward cash flow of the hydroelectric development in Example 5.5 may be altered, first by the effect of inflation as shown in Example 8.1, and secondly by the use of a loan.

EXAMPLE 8.4

A twenty-year loan has been negotiated for the hydroelectric development in Example 5.5.   The terms provide that the debt shall be redeemed over a period of 15 years by making equal payments at six-monthly intervals.   These payments shall cover principal and interest and commence in year 6.   What project costs will the owner have to meet out of his own resources, which include revenue earned by the project, over the first 30 years?

In Example 8.1 the total debt to be repaid at the start of year 6 was assessed as £47 938 000.   To meet the loan terms as defined, 30 six-monthly payments will be required, calculated at a half-yearly interest rate of 3%, each amounting to:

$$£47\ 938\ 000 \times \frac{0 \cdot 03}{1-(1 \cdot 03)^{-30}} = £2\ 446\ 000$$

The annual capital charges will then be twice this sum, namely,
    £4 892 000

Using the cash flow data from Example 5.5 it is now possible to derive the costs of the project which have to be met from revenue or other sources.   These are set out in column *F* of Table 8.4.

The figures in column *F* of Table 8.4 show the amounts that, over the whole 30 years, would have to be recovered from revenue in order that the project should break-even should no other initial financing be available.   However, additional finance will certainly be needed in years 1–5 before any revenue is earned.   This will be provided by further borrowing, by the issue of share capital (in the case of a company), or from

121

TABLE 8.4

A Year	B Expenditure cash flow	C	D Expenditure met by loan £ × 10³	E Loan capital charges £ × 10³	F Net project costs (C + D + E) £ × 10³
	at year 1 prices £ × 10³	after 3% inflation £ × 10³			
1	5 050	5 126	−3 845		1281
2	12 650	13 228	−9 917		3311
3	15 100	16 258	−12 194		4064
4	12 650	14 030	−10 522		3508
5	5 050	5 769	−4 327		1442
6	84	99		4890	4989
7	1 484	1 826		4890	6716
8	96	120		4890	5010
9	96	123		4890	5013
10	1 596	2 150		4890	7040
11	108	147		4890	5037
12	108	152		4890	5042
13	1 608	2 360		4890	7250
14	120	178		4890	5068
15	120	184		4890	5074
16	120	189		4890	5079
17	120	195		4890	5085
18	120	201		4890	5091
19	120	207		4890	5097
20	120	213		4890	5103
21	120	219			219
22	120	226			226
23	120	233			233
24	120	240			240
25	120	247			247
26	120	254			254
27	120	262			262
28	120	270			270
29	120	278			278
30	120	286			286

finance that is already available from earlier borrowing (or issue of shares) or from past earnings in the form of surplus cash flows not yet re-invested or distributed to the owners. If revenue is to cover all costs, this further financing will have to be recouped out of later revenue together with an addition for interest sufficient to justify its use for this project. The complete financial plan requires, therefore, a more extended statement than that shown in Table 8·4, in the form of an overall budgetary statement, as shown in Example 8·5.

BUDGETARY PRESENTATION

The comprehensive budgetary statement would normally be prepared jointly by the engineer and the accountant and would indicate:

(a) the amount of cash which has to be provided year by year to meet any deficits which may arise from the project, or the surplus cash that will be released for some other use, as the case may be

(b) the impact of the project upon the net profit or loss of the undertaking, as assessed under normal accounting conventions.

Such a presentation would, among other things, provide the lender with the evidence that the projected revenues of the undertaking would be on such a scale as to provide for the payment of interest on, and the eventual repayment of, the loan; or, if this was not the case, that other adequate financing arrangements had been made for these purposes.

Example 8.5 gives a budget presentation for a major port development.

EXAMPLE 8.5

It is assumed:

(1) that all capital expenditure will be financed by borrowing at 8% per annum, that interest during the construction period will be added to the amount of the loan, and that repayment will be made in 20 equal annual instalments covering capital and interest, the first falling due at the end of year 6

(2) that for accounting purposes depreciation is calculated at 2% per annum, on a straight line basis on the capital expenditure incurred up to the end of year 5

(3) that it will be possible to invest idle funds or borrow to meet small additional requirements at 7% per annum.

Table 8·5 represents the finance budget over the first fourteen years of the project, while Table 8.6 sets out the net accounting income over the same period.

It will be noted that the operating income and expenditure figures (columns P and K of Table 8.6) constitute the operating cash receipts and payments (columns G and F of Table 8.5) subject to a varying time displacement. The difference arises because the figures used for profit

123

TABLE 8.5. FINANCE BUDGET

Year	A Loan at beginning of year: (D-E) in previous year £×10³	B Capital expenditure at year end £×10³	C Interest at 8%: (A×0·08) £×10³	D Sub-total: (A+B+C) £×10³	E Repayment £×10³	F Operating cash payments £×10³	G Operating cash receipts £×10³	H Interest on additional finance at 7%: J in previous year ×0·07 £×10³	I Additional finance required (−) or available (+): G−(E+F+H) £×10³	J Cumulative additional finance required (−) or available (+): J in previous year+I £×10³
1	—	8 000	—	8 000	—	—	—	—	—	—
2	8 000	10 000	640	18 640	—	—	—	—	—	—
3	18 640	12 000	1491	32 131	—	—	—	—	—	—
4	32 131	4 000	2570	38 701	—	225	900	—	675 (+)	675 (+)
5	38 701	1 000	3096	42 797	—	475	1900	47 (+)	1472 (+)	2147 (+)
6	42 797	—	3424	46 221	4359	545	2000	150 (−)	2754 (−)	607 (−)
7	41 862	—	3349	45 211	4359	595	2900	42 (−)	2096 (−)	2703 (−)
8	40 852	—	3268	44 120	4359	645	3900	189 (−)	1293 (−)	3996 (−)
9	39 761	—	3181	42 942	4359	695	4900	280 (−)	434 (−)	4430 (−)
10	38 583	—	3087	41 670	4359	745	5900	310 (−)	486 (+)	3944 (−)
11	37 311	—	2985	40 296	4359	795	6900	276 (−)	1470 (+)	2474 (−)
12	35 937	—	2875	38 812	4359	800	7450	173 (−)	2118 (+)	356 (−)
13	34 453	—	2756	37 209	4359	800	7500	25 (−)	2316 (+)	1960 (+)
14	32 850	—	2628	35 478	4359	800	7500	137 (+)	2478 (+)	4438 (+)

TABLE 8.6. BUDGET OF NET ACCOUNTING INCOME OR EXPENDITURE

Year	K Operating expenditure £ × 10³	L Depreciation £ × 10³	M Interest on fixed loan: from C £ × 10³	N Interest on additional finance: from H £ × 10³	O Total expenditure: K+L+M+N £ × 10³	P Operating income £ × 10³	Q Net income (+) or expenditure (−) P−O £ × 10³
1	—	—	—	—	—	—	—
2	—	—	640	—	640	—	640 (−)
3	—	—	1491	—	1491	—	1491 (−)
4	250	600	2570	—	3420	1000	2420 (−)
5	500	680	3096	47 (+)	4229	2000	2229 (−)
6	550	700	3424	150 (+)	4524	2000	2524 (−)
7	600	700	3349	42 (−)	4691	3000	1691 (−)
8	650	700	3268	189 (−)	4807	4000	807 (−)
9	700	700	3181	280 (−)	4861	5000	139 (+)
10	750	700	3087	310 (−)	4847	6000	1153 (+)
11	800	700	2985	276 (−)	4761	7000	2239 (+)
12	800	700	2875	173 (−)	4548	7500	2952 (+)
13	800	700	2756	25 (−)	4281	7500	3219 (+)
14	800	700	2628	137 (+)	3991	7500	3509 (+)

calculation in Table 8.6 are amounts due and payable, or receivable, in respect of the services actually performed during each year; there will normally be a short delay between the performance of the service and the settlement in cash.   The difference between the two sets of figures is often known as a change in working capital.

The figures given in column *J* of Table 8.5 indicate that implementation of the project when combined with the loan arrangement described under assumption (1), will:

(a) leave the cash position of the undertaking unchanged in years 1 to 3

(b) generate a cash surplus in years 4 and 5

(c) create a need for additional finance in year 6; this need will increase up to year 9 after which it will decline and be extinguished by year 13.

For the reasons given in Chapter 3 there is no precise relation between the income shown in Table 8.6 and the cash surplus in Table 8.5.

ANNUAL COSTS

As stated in Chapter 2, it is sometimes useful to present the capital costs of a project in the form of an annuity.

EXAMPLE 8.6

The following data relate to a hydroelectric project of the type described in Example 5.5:

(a) the total investment, including interest during construction, is £60 million, represented by two classes of assets costing £52·5 million and £7·5 million, with effective lives of 60 years and 30 years respectively

(b) the interest rate to be used is 6% per annum

(c) annual running expenses are constant at £120 000

(d) the annual output from the project is constant at 2100 million kWh.

On the foregoing basis, the costs can be expressed in annual terms as:

	£
annuity equivalent of £52·5 million over a life of 60 years*	3 248 500
annuity equivalent of £7·5 million over a life of 30 years*	544 800
running expenses	120 000
	3 913 300

---

* The annuity A is given by using the inverse of the formula of Appendix F, Table 3:

$$A = P \frac{r}{1-(1+r)^{-n}}$$

where P is the value of the initial investment at time 0, n is number of years' life, and r is the annual rate of interest. In practice the initial investment may be spread over several years; the various cash payments that comprise it must therefore be converted to a single value at time 0 by using for each payment the formulae of Appendix F, Table 1 or 2.

The algebraic relationships given in Appendix F show that the following relation holds:

annuity equivalent = sinking fund + interest on initial capital sum

Thus, taking the capital investment of £52·5 million:

	£
6% on £52·5 million	3 150 000
sinking fund to redeem £52·5 in 60 years at 6% (Appendix F, Table 4)	98 500
annuity equivalent	3 248 500

The latter figure can be obtained directly from the formula given at the beginning of this note.

This annual amount can be used as a guide in judging comparative costs, and in pricing, by dividing it by the annual output of 2100 million kWh to give an average cost of £0·001 86 per kWh (or 0·186 new pence).

Such a calculation must be used with caution in view of the implicit assumptions on which it is based. It assumes that the amount of power sold annually is constant over the whole of the active life of the undertaking. It assumes that at the end of the first 30 years the plant and equipment comprising the initial investment of £7·5 million will be taken out of service and replaced by similar assets at the same cost. Such an assumption takes no account of either technological or economic changes which, over a period of 30 years, may make such a replacement undesirable.

While for these reasons such calculations must be used with care, they are nevertheless useful in preliminary studies. However, it is always desirable in the later stages of a project to lay out the financial plan in detail as shown in Table 8.5. If there should be unusually large financial needs at particular points of time special planning may be necessary.

VARIABLE COSTS

Although the actual price at which a commodity or service is sold will be conditioned by demand, by competition, by the policy of the undertaking, and perhaps by government policy, it must be considered in relation to its cost of production. In the long run if the enterprise is to be successful the cost of production must be covered by revenue earned (unless there is a government subsidy).

It is often convenient to divide this cost of production into:
  (a) fixed costs
  (b) variable costs, in some contexts called running costs.

The fixed costs are those which are independent of output, whether of goods or services over a given range.

The variable costs are dependent upon use or output. Under this heading would usually be included consumable stores or raw materials used in production, the cost of maintenance, and a proportion of the expenditure on salaries, wages, rates and taxes, insurance. Some part of the latter will usually be approximately constant over a range of output and will therefore be classified as fixed. This would apply to such items as Head Office expenses, insurance, rates and taxes etc.

An early example of the application of these concepts is to be found in the method of analysis used by the Electricity Commissioners in the 1920s to determine the cost of producing electricity from the British power stations of the day. The various costs were divided into two groups, those directly proportional to the amount of electricity generated, and those which remained constant regardless of output within the capacity of the stations. This allowed the cost of electricity from a steam power station to be expressed in the simple form

$£X$ per year per kW of effective capacity together with Y pence per kWh.

In this way the cost of producing electricity was determined in relation to the use made of the plant, that is, in terms of the annual load factor, the average cost per unit falling as the load factor rose.

While more sophisticated methods of analysis have been developed in recent years, the basic principle of expressing costs in terms of fixed and incremental (or marginal) components demonstrates the economic advantage of securing a high utilization from a fixed investment. The economic benefit to be derived from high utilization becomes of greater significance as the ratio of fixed to variable cost components increases. Nuclear power stations offer a well known example.

In an analysis of this kind, matters to be kept in mind include the following. The fixed costs are largely determined once the total investment is known and the financing terms have been agreed; moreover, once incurred they are immune to the effects of inflation. They are, however, only fixed for the life of the asset. In a long term analysis all costs tend to become variable. Variable costs may tend to increase with time, partly because more maintenance is needed as physical assets deteriorate with increasing age. They may also rise if the real cost of wages in relation to general price levels rises.

These observations relate particularly to those projects in which the annual costs are dependent to a material degree on the use which is made of the assets. In some types of engineering works however, such as bridges, the total annual costs are virtually fixed regardless of whether the works are used or not. In the case of a steel railway bridge for example, the cost of painting, inspection, etc., does not depend significantly upon the traffic using the bridge and these expenses would be reduced hardly at all in the event of it being closed. In such cases it may be said that once the work has been completed, all the costs become fixed during its economic lifetime, that is, until it pays to replace it. This is substantially true of a hydroelectric project where the annual costs are but little affected by use.

MULTI-PURPOSE PROJECTS: ALLOCATION OF COSTS

It is becoming increasingly common for large engineering projects to be planned to serve more than one principal purpose. To such projects the term 'multi-purpose' may be applied—a term which was first used in the 1930s in connexion with schemes of river basin development where the primary benefits sought usually included irrigation, flood control, navigation, and power.

While the principles to be applied to the economic evaluation of a multi-purpose project are similar to those which have already been described in connexion with single-purpose projects, the evaluation of a multi-purpose project presents special problems concerning the allocation of project costs among the various participants who will share in the benefits.

For the purposes of the discussion which follows, it is assumed that engineering studies have led to the conclusion that the project as a whole can be justified on economic grounds and that these studies have been elaborated with the object of securing from the project the greatest surplus of benefits over costs in terms of their present values. It is further assumed that these studies make it possible to isolate those portions of the total costs which are to be incurred solely in the interests of a particular participant—as for instance the inclusion of a power station in a project primarily designed for flood control.

The isolation of these 'separable costs' as they may be called leaves the balance of the project costs, the 'common costs', to be divided among the

129

participants in some equitable manner. In the final analysis this must be a matter of mutual agreement. In practice, therefore, the relative bargaining power of the participants may affect the result.

In considering how the costs may be divided, it is important to observe that those assigned to a particular participant should not be greater than those which would be incurred if the benefits were to be obtained from the most favourable independent alternative source. Thus the costs allocated to the production of electric power should be limited by the cost of obtaining an equivalent supply from another source, such as a thermal power station, independent of the project.

One method of allocation is the 'separable costs–remaining benefits' method. The amounts allocated to each purpose are in proportion to the remaining benefits from each purpose after subtracting from each its separable cost. Each benefit in this context is the single-purpose benefit or, if lower, the alternative cost of providing that benefit independently.

The foregoing procedure can best be illustrated by means of a simple example of a multi-purpose river control project costing £20 million.

EXAMPLE 8.7

The project will provide benefits in the form of increased agricultural output as the result of irrigation and flood control, and of reduced transportation costs arising from improved river navigation. Other benefits will arise, such as improved fishing and recreational facilities, but these are ignored because of their lesser significance coupled with difficulty of valuation.

If £8 million of the project cost is specifically due to irrigation and flood control (in the sense that had the benefits from this not been wanted the £8 million could have been saved on the project) and £5 million similarly to navigation, the remaining £7 million (or common costs) has to be divided between these purposes. How can this be done in an equitable manner, given that:

(a) the increase in the present value of agricultural benefits is set at £12 million

(b) the capitalized cost (present value) of alternative transportation to the navigation works is set at £6 million

(c) navigational benefits downstream of the works accrue independently of the proposed navigation works and are expected when capitalized in present value terms to amount to £6 million?

Take first the surpluses of benefits over their respective separable costs. It is assumed that the cost of providing these benefits by carrying out an independent project would exceed those benefits. The surplus from agriculture is

£12·0 million − £8·0 million = £4·0 million

that from transportation is

£6·0 million + £6·0 million − £5·0 million = £7·0 million

In the simple example taken it might well be considered reasonable to divide the common costs in proportion to the above surpluses.

Thus £7 million of common costs would then be allocated between agriculture and transportation in the ratio of 4 to 7, or £2·55 million and £4·45 million respectively.

The total cost allocation can then be summarized as follows.

	Agricultural	Transportation	Total
Separable costs: £ × 10⁶	8·00	5·00	13·00
Common costs: £ × 10⁶	2·55	4·45	7·00
	10·55	9·45	20·00

There are various other methods which might in some circumstances form a basis for negotiating cost allocations when agreement on the 'separable costs–remaining benefits' method is either inappropriate or unattainable.

MARKETABILITY METHOD

As the title suggests, this method involves the apportionment of common costs in the ratio of the revenues that would be received for the services provided if sold at market prices. The method is often difficult to apply in practice, since such services as flood control or navigation are not readily marketable.

BENEFIT METHOD

Under this method, common costs are apportioned in the ratio of the estimated economic values of the different benefits to be derived from the proposed development. It is fundamentally similar to the marketability method since the value of benefits is often determined by the market price which can be attached to them. However, it also allows for the possibility that some of the benefits may not be traded in any market. The practical application of this method is often complicated by the difficulty of evaluating the benefits, some of which may be indirect or intangible.

USE OF FACILITIES METHOD

This method involves the allocation of common costs in proportion to the use made of a facility which is jointly shared by the various participants. The method assumes that a unit of measurement can be found which can be applied to all the users; thus, in some cases, a cubic metre of water might serve as a common unit in considering the apportionment of costs, but this would be inapplicable in the case of a power project where not only quantity but head is an essential feature.

SPECIAL COST METHOD

This envisages the apportionment of common costs in proportion to the special costs incurred on behalf of each of the participants. Although simple, the procedure may be rejected because it takes no account of the relative benefits accruing to each of the participants by reason of the common facility.

ALTERNATIVE SINGLE-PURPOSE EXPENDITURE METHOD

This method involves the division of the common costs in proportion to the hypothetical alternative expenditures which would be incurred if each of the participants were separately to construct a facility to provide the benefits which they would receive from the joint project. It may sometimes be difficult to assess with accuracy the cost of a hypothetical alternative single-purpose development, but usually fewer uncertainties arise than with some of the alternative methods mentioned above.

Whatever the method chosen it would be difficult to justify an allocation of common costs which led to the service in question costing more than it would do if the best alternative were used.

Appendix A   Alternative methods of investment appraisal

The methods of investment appraisal discussed in this handbook use compound interest calculations for the comparison of costs and benefits arising at different times.   Other types of calculation have been, and to a considerable extent still are, used for this purpose.   These methods have inherent weaknesses absent in the discounted cash flow approach, but the two most common are discussed below.   The first is based on the payback period.

PAYBACK PERIOD

This may be defined as the length of time which elapses between the inception of a project and the time when the cash benefits flowing from it have equalled the initial outlay.   A project is accepted if it has a payback period less than some predetermined standard, say 5 years.   If projects have to be ranked, the project with the shortest payback period that also satisfies the chosen standard is accepted.

It can be shown from an example that this method can lead to a different ranking of projects from that produced by the present value method. Consider projects P and Q in Example 2.5.

Project P has a payback period of $2\frac{1}{6}$ years, for the initial outlay is equal to the receipts for the first two years plus one-sixth of the receipts for the third year: $£3000 + 5000 + (\frac{1}{6} \times 6000) = £9000$.   Similarly project Q has a payback period of $1\frac{3}{4}$ years: $£6000 + (\frac{3}{4} \times 4000) = £9000$.   On this basis project Q is shown to be preferable to project P contrary to the results of the net present value calculation.

The defects of the payback method are that it does not take into account the whole of the economic lives of the projects being appraised and that any cash flow which arises after the end of the payback period has no effect on the calculations.   A project may have a very short payback period and yet have no cash receipts after this period expires; in this event the cash receipts will not cover the interest cost associated with the project.   On the other hand, a project with high cash benefits after the end of the payback period will be given no credit for these.   This is why project P, shown to be the better project in Example 2.5, is rejected by the payback test. Again, the payback period method takes no account of the different timing of net cash receipts within the payback period.   Two projects may have the same payback periods but the receipts of one may arise earlier on the average than those of the other.

RATE OF RETURN ON BOOK VALUE

The second method of investment appraisal is based on the expression of accounting profit as a percentage return on the capital invested, as measured by accounting book values; the resulting percentage is then compared with some rate which represents a standard of acceptability.   There are many possible variations of this method; it will be illustrated here by an example which expresses the average accounting profit over the lifetime of a project as a percentage of the average book value of the assets.

Consider a project that requires an initial outlay of £8000 and is expected to earn net cash receipts of £1000, £3000 and £6000 at the end of 1, 2 and 3 years.   The outlay is for the purchase of a machine which will be worn out at the end of year three; it will then have no scrap value.   The required return is 10% p.a.   The net present value of the project, based on the method used in this handbook, is in £s:

$$-8000+\frac{1000}{1\cdot10}+\frac{3000}{(1\cdot10)^2}+\frac{6000}{(1\cdot10)^3} = -8000+909+2479+4508$$
$$= -104$$

Thus on a net present value calculation the machine should not be purchased because it will not provide the required return on the capital invested.

However, if it is assumed that for the calculation of accounting profit the machine will be depreciated on a straight-line basis, and that there is no lag between the cash flows becoming due and being actually received, the calculation of the net accounting profit from the project would be as shown in Table A.1.

134

TABLE A.1

Year	Revenue	Depreciation of machine	Net profit or loss	Book value of machine		
				at start	at end	average
1	1000	2667	−1667	8000	5333	6 667
2	3000	2666	334	5333	2667	4 000
3	6000	2667	3333	2667	0	1 333
Total			2000			12 000
Average per annum			667			4 000

Average return on capital based on book value = 667/4000 = 16·7%

If a manager were to rely on this method of appraisal, he would purchase the machine because the calculated average return of 16·7% is greater than the required standard of 10%. He would find, however, that if he borrowed the initial £8000 at 10% p.a. and repaid the loan with interest as the revenue came in, there would be a deficit at the end of the three years.

There are various alternative ways of depreciating assets for accounting purposes. In general, none of those normally in use will give the same result as that obtained by the discounted cash flow approach, though it is possible to devise an accounting method that will give such a result.

10

Appendix B   Sources of finance in the private sector

In this appendix a short survey is given of some of the more important sources of finance, and of types of financial obligation, in the private sector of the economy. The survey is not exhaustive and further reading is suggested in the final paragraph. The various types of finance described relate to the requirements of a company incorporated with limited liability, this being the most important form of business organization in the private sector of the economy.

LONG-TERM LOANS

Various names are given to loans raised by companies for periods of several years: Debenture Stocks, Loan Stocks, and Notes are the most common. When such loans are raised or 'issued' the terms of issue state the interest to be paid and indicate when repayment will be made. The date of repayment may be specified, or it may be selected by the company within a specified period. Sometimes the loan may be perpetual, that is, having no specified date for repayment; it is then repayable if the company is wound up or goes into liquidation (i.e. comes to an end).

Loans will usually be issued in units having a given nominal or par value, for example £1 or £100, each unit being separately transferable so that the holder can easily sell it if he wishes. The holders of units are paid a fixed amount of interest, calculated as a percentage of the nominal value of their holding. Interest is normally paid at half-yearly intervals. A loan stock may be issued and repaid at par, that is at its nominal value, in which case a subscriber for £100 of loan stock would pay the company £100 and he (or the current holder) would be entitled to receive £100 when the loan was repaid. But other provisions are possible. A loan may be issued at a premium (a price greater than the nominal value) or at a discount (a price below the nominal value); and similarly repayment may be made at a premium or at a discount. If a stock is not issued and repaid at par, its effective interest cost will differ from the fixed rate of interest expressed to be payable on the nominal value. Thus, if a stock unit of £100 is issued at £90 (a discount of 10%) and is repayable in twenty years' time at par, and its nominal rate of interest is 6% payable annually, its effective interest cost is higher than 6% p.a. and is given by the value of r in the equation

$$90 = \frac{6}{1+r} + \frac{6}{(1+r)^2} + \cdots \frac{106}{(1+r)^{20}}$$

By issuing a stock at a premium or a discount the effective rate of interest can be varied at the time of issue according to the market conditions then ruling. For technical reasons this may be more convenient than altering the nominal rate of interest.*

The interest on a loan stock is fixed by a legal contract under which a company is liable to pay the interest whether it makes a profit in any given year or not. The holder of a loan stock is often given some form of security for the payment of interest and repayment of the loan. One type of security takes the form of a mortgage on assets of the company, usually on land and buildings. In this event the holders can have the assets concerned sold if the company fails to abide by the conditions under which the stock was issued. They are then repaid the amounts due to them out of the proceeds (to the extent that these are sufficient) without prejudice to their right to claim any balance unpaid from the general assets of the company. If a company's assets are subject to a mortgage, it cannot sell them, or alter them materially, without obtaining the consent of the lenders. The loan will then probably be called a mortgage debenture, or a debenture carrying a fixed charge.

It is also quite common for all the assets of a company, subject possibly to some specific exclusions, to be charged as security for a loan under an arrangement known as a floating charge. This arrangement enables the company to deal with its assets in the normal course of business unless, and until, default is made in the terms of the loan, in which event the lenders have the right to appoint a receiver whose function it is to take over the assets and sell them or otherwise deal with them in a way which

---

* Issue and redemption at a premium or discount may have tax consequences, both for lender and borrower. Specialist advice should therefore, as always, be sought before decisions are taken.

will enable him to repay the amounts due to the lenders. Such a loan will usually be called a debenture carrying a floating charge.

The type of security, if any, which is given by the company will influence the risk borne by the lender. A lender whose loan is secured by the mortgage of an asset that is unlikely to lose its value will bear a very small risk indeed; the holder of a floating charge will bear a somewhat greater risk, and a lender without security an even greater one. The greater the risk the greater the interest rate which will normally be required to induce an individual to invest. One risk which is invariably present in such transactions is the risk of inflation. Since the interest cost of a loan stock is fixed in actual money terms, the higher the rate of inflation the lower the interest cost to the company when measured in units of constant purchasing power and the lower the real interest earned by the lender. The awareness of the risk may be expected to induce lenders to demand a higher contractual interest return in compensation. However, if the company expects a higher rate of inflation than lenders in the market, the issuing of loan stocks may seem to be a means of raising capital at a low real interest cost.

As noted in Chapter 3, the effective interest cost of loan capital to a company is reduced by taxation because loan interest paid is allowed as a deduction in computing the amount of profit on which corporation tax is assessed. Assume that a 6% loan stock is issued and repayable at par, and thus has an interest cost of 6% p.a. Assume also that the company's rate of tax is 40%. Then the loan has an effective net cost of 3·6% (2·4% being recovered in the form of relief from taxation). If inflation is expected at the rate of 3% p.a., the real interest cost of the loan is eventually reduced almost to zero. On the other hand, the higher the proportion of total finance raised in the form of loans, the greater is the risk that a period of adverse trade, leading to inability to meet the contractual rights of the loan holders, will cause the ordinary shareholders permanently to lose their control of the company as the result of loan holders enforcing their legal rights. The risk to the loan holders themselves also rises as the proportion of loan finance to total finance rises. These considerations limit the amount of loan finance that can be raised.

Interest rates required by lenders tend to rise and fall according to changes in the condition of the national economy. Suppose a stock is issued and repayable (in 20 years' time) at par and bears a nominal interest rate of

5% p.a. As long as interest rates remain more or less constant, and as long as there is no reason to doubt the ability of the borrower to meet his obligations, the stock can be expected to change hands in the market at a price close to par. However, if interest rates in general rise, so that the rate payable on stocks of this kind becomes, say, 7%, purchasers of this stock will, in view of the alternatives available to them, no longer be willing to pay so high a price for it. The market price will tend to fall to

$$P = \frac{5}{(1 \cdot 07)} + \frac{5}{(1 \cdot 07)^2} + \cdots \frac{105}{(1 \cdot 07)^{20}}$$

A company can estimate the effective current interest cost of a given type of loan stock by observing the market prices of similar stocks and calculating the effective rate of interest associated with such stocks at these prices. Such a rate is sometimes known as the redemption yield of the stock in question.

INSTALMENT FINANCE

Instalment finance is often thought of in relation to domestic purchases. In fact a good deal of instalment business is concerned with the pricing of a wide range of industrial equipment and is thus a source of medium to long-term finance for companies.

SHORT-TERM LOAN FINANCE

There are several ways in which a company can supplement its long-term finance for short periods. Perhaps the most important is by borrowing from a bank on overdraft, a facility on which interest is charged at a rate which is usually one or two per cent above Bank Rate (the rate at which the Bank of England will lend money for short periods to accredited customers on first-class security). In the past, British banks have generally preferred to restrict their lending to advances to meet short-term requirements, such for example as arise from seasonal fluctuations in trade, or from the need for a bridging operation which may occur during the construction of an asset, pending the acquisition of finance of a more permanent nature. Some longer term loans are in fact made, but it is generally banking policy in Britain not to provide permanent finance for a business.

Finance may be provided by trade credit. A person provides finance to a business if he gives it credit by waiting for payment after he has supplied it with goods or services. Such finance may however have a cost to the business receiving it, for example because the delay in payment involves losing a cash discount, or buying at a higher price. Moreover, undue delay in settlement may injure the buyer's reputation and hence the willingness of suppliers to continue to deal with him. Trade credit is however normal in many types of business, the period of credit allowed being usually determined by the custom of the particular trade.

OWNERSHIP FINANCE: PREFERENCE SHARES

Preference shares have some features in common with loan stocks. They have a nominal value on which a fixed rate of interest (or dividend as it is more usually called in this case) is paid. They may be redeemable or irredeemable and they may be issued or repaid at a premium (there are legal restrictions on their being issued at a discount). Such shares are, however, usually transferable in units of £1 or less.

However, there are important differences between preference shares and loan stocks. No dividend can be paid to ordinary shareholders unless the preference share dividend has been paid in full. No dividend can be paid to holders of preference shares unless the company makes sufficient profit in that year or has available undistributed profits accumulated from previous years; and even then the company is not obliged to pay a dividend if its directors consider it inadvisable. There may or may not be provision for shares to be cumulative—that is for a short-fall in a dividend payment in one year to be made up in a later year. It is not usual for any assets to be pledged as security for preference shareholders. Preference shares can only be repaid if the company is wound up; or if an equivalent amount of new capital is issued at the time of repayment; or if an equivalent amount of the accumulated profits of the company is 'frozen' so that it cannot be distributed. The dividend paid to preference shareholders is not allowable as a deduction in computing the taxable profit of the company, and this disadvantage in terms of taxation, as compared with loan finance, probably accounts for the fact that preference shares are rarely issued nowadays.

OWNERSHIP FINANCE: ORDINARY SHARES

The holders of ordinary shares (sometimes called equity shares because they carry the residual interest in the company's assets) bear the greatest uncertainty of all those who provide finance regarding the return which they will receive. Ordinary shares have a nominal value; they can be issued at a premium over this value but not normally at a discount. The amount raised when the shares are issued cannot be repaid unless the permission of the court is obtained or the company is wound up. No dividend can be paid to ordinary shareholders unless the company has made a sufficient profit to cover the dividend after deducting interest on loans and after paying any preference dividend which may be due*; but subject to this there is no upper limit to the dividend which may be paid. Similarly, if the company is wound up, the ordinary shareholders may be repaid whatever remains after all liabilities have been paid and usually after the preference shareholders have been repaid. Thus the ordinary shareholders might be described as residuary beneficiaries: they are entitled to whatever is left after everyone else has had his due and they take the chance that this may be much or little.

The formula used above for the calculation of the interest cost of loan stock may be regarded as a starting point for the estimation of the rate of return which ordinary shareholders require on finance provided by them, which in turn is relevant in selecting investment projects within the company. If it were known what future dividends were expected by the holders of ordinary shares already issued by the company, the required rate of return could be ascertained readily enough. If the current market price per share is P and the expected dividend per share in year n is d_n, then the rate of return which shareholders are implicitly requiring (which may be called the cost of equity capital) can be determined as r in the equation

$$P = \sum_{n=1}^{n=\infty} \frac{d_n}{(1+r)^n}$$

A major difficulty is that of estimating the level of dividends expected in the future. A plausible approach is perhaps to assume that shareholders generally arrive at their future dividend expectations by a process of extrapolation, based on the company's dividend record in the recent past. Thus, if the dividend has been stable for some years, it may be expected to remain so; if a steady rate of growth has been observed this may be expected to continue; and so on. If there are special reasons for expecting a new dividend pattern in future, this can be allowed for. This in fact

---

* The determination of profit for this purpose is based on the accounting conventions referred to in Chapters 3 and 8. These conventions are founded to a considerable extent on rules laid down by judges in cases in which the amount of profit that could be distributed was in dispute.

141

appears to be the way in which buyers of shares tend to behave, and calculations may be based on this assumption.

It is likely that the required rate of return on any new equity capital raised would be close to the current figure, provided that the amount of new capital raised was not an unduly large proportion of the total capital employed by the company and that no significant changes in the character of the company's business had occurred.

RETAINED EARNINGS

It is unusual for companies to pay out all the cash generated by their activities as a dividend each year; some will be retained for re-investment to replace the assets that are depreciating in value, and some will usually be invested to expand the business of the company. These retentions are an additional source of equity capital for if they were not used in this way they could, subject in some cases to the observance of some legal formalities, be paid out to the ordinary shareholders. This suggests that the rate of return that should be sought on such retentions should be the same as that required on the investment of newly raised equity capital. There are, however, two qualifications. First, the raising of new equity capital involves considerable expense in the form of advertising, legal fees, commissions, and so on; this raises its effective cost. Secondly, additional tax is payable on profits distributed as dividends. For these reasons the amount of retained cash needed to produce a given future cash flow is less than the amount of new equity capital that would be needed to produce the same flow.

For a more detailed study of the various sources of finance and types of financial obligation see PAISH F. W. *Business finance*. Pitman, 1965; and MERRETT A. J. and SYKES A. *Capital budgeting and company finance*. Longmans, 1965, p. 32.

Appendix C   Development planning

It has been suggested* that 'next to the maintenance of world peace . . . raising the standards of living of the population of the underdeveloped countries . . . is the most important problem of our time'.

It is perhaps due to the acceptance by the richer countries of the obligations implicit in such a view, that engineers have been increasingly directing their attention to those regions which have remained relatively untouched by the industrial developments of the past 200 years. Such regions are usually characterized by a subsistence agricultural economy, often supported by the export of specialized plantation produce, and in some cases by minerals. To some engineers these countries offer a special appeal because of their great potential and the possibilities they present for planning economic development.

During the early stages of development† such countries tend to be characterized by:

- (a) a wide range between the incomes enjoyed by the rural population and by the more prosperous classes: traders, administrators, landowners
- (b) a high illiteracy rate arising from inadequate education facilities
- (c) a high and largely concealed unemployment combined with ineffective farming methods and low productivity
- (d) poor communications and the isolation of the rural population
- (e) a dearth of factual information and reliable statistics.

---

* MACBEAN A. I. *Export instability and economic development.* Allen and Unwin, 1967.
† It is important to avoid dogmatism in discussions of these matters. The histories of Australia, New Zealand, and Denmark, for example, show that it is possible for a high standard of life to be founded on a prosperous agriculture, the products of which are exported.

ECONOMIC DEVELOPMENT

In countries which have had a long history of industrialization, continued economic growth depends upon the application of special skills and technological research to the conversion of raw materials, whether indigenous or imported, and to the maintenance of a competitive position in overseas markets. Under such conditions steady improvement in industrial processes reduces the direct labour content of many products and, by increasing the productivity of the individual, makes it possible for a given body of labour to cover an ever wider economic field.

In the underdeveloped countries, the immediate problem of the development planner is usually to raise the productivity of agriculture and provide work for the underemployed.

In regions where little industry has been developed, it is usual to find that considerably more than half the population* is employed on the land as was commonly the case in Europe in the nineteenth century. In the United States for instance, the farm population in 1900, before machines had been applied to agriculture, was 40% of the total. Thirty years later this had fallen to 26% while in 1968 less than 5% of the employed population gained a livelihood on the land; and this at a time when the country was producing massive food surpluses. But industrialization is only one aspect of economic development. It does not follow that in a particular case, plans for development should concentrate on industrial aspects. The most important may be directed towards increasing the production of food.

Such increase may be achieved by the adoption of better agricultural practices, by the use of fertilizers, by irrigation, and by the use of machinery. But above all, success depends upon the education of the farmers and the degree to which they can be persuaded when necessary to abandon their traditional ways in favour of better methods.

The development of agriculture should be related to plans for industrialization. Except to the extent that income can be obtained by exporting, and this is an important qualification, no significant demand for manufactured goods can arise until the rural population can produce more food than is required for their own subsistence.

---

* Statistics published by the OECD indicate that in the case of Turkey the proportion of the population employed in agriculture in 1965 was 75%.

INDUSTRIALIZATION

It has been pointed out* that economic growth has tended in the past to be accompanied by a continuous decline in the importance of agriculture relative to other sources of employment; thus in many countries industry is continually recruiting labour from the countryside. Experience shows that when labour first enters industry from agriculture, its productivity is often low compared with that of labour which has become adapted to industry over a long period of time. For this reason labour costs in newly-established industries are often unexpectedly high in spite of the relatively low wages paid, and considerable time may be required for agricultural labour to adjust itself to the discipline of factory life.

Industrialization usually forms an integral part of planned overall economic development. The types of industry to which priority should be given depend upon the kinds of natural wealth awaiting development and especially upon the existence of minerals, such as oil, offering opportunities for profitable exploitation. Whatever the broad pattern of development envisaged, it will usually be one of the primary objectives of an industrialization policy to replace some imports by locally manufactured goods.

The success of a development programme may depend upon the choice of industries and the sequence in which their development takes place. It is argued that these early industries should be directed to meeting the more fundamental needs of the local market, in so far as is consistent with utilizing the natural resources of the country such as its agriculture, fisheries and forests and those minerals which can be more readily processed.

Among such industries might be cement, ceramics, textiles, furniture, the canning of food, brewing, and the preparation of hides and skins. At a somewhat later stage the production of pulp and paper might be undertaken, using indigenous sources of cellulose if available or imported pulp.

Desirable characteristics of such industries are: a modest capital requirement in relation to the value of the product, the comparative ease with which the necessary skills can be acquired, and the broadly based nature of the markets which can be developed. Another desirable feature is that economic production should be less dependent upon the scale of activity

---

* LEWIS W. A. *Theory of economic growth.* Allen and Unwin, 1955.

than is the case with more complex industries using elaborate and costly equipment.

Again, however, it is important to avoid dogmatism. Some of the examples of most rapid economic growth have been regions such as Malaya, Hong Kong, South Africa and Brazil, where intensive early development took place without the assistance of centralized planning.*

As people become more prosperous in a steadily expanding economy, they tend to spend a larger proportion of their income on manufactured goods and services and less on food. Thus the proportion of the national income which is spent on agriculture tends to decline while a greater proportion is spent on industry, urban housing, services, etc.

Eventually, in a highly-developed economy, a stage may be reached when employment in the services sector will exceed that offered by agriculture and industry combined.† This stage was reached in the USA by 1960 and it has been confidently predicted that it will be attained in a number of Western European countries in the 1970's. It does not, however, follow that the same pattern will be repeated in every country. A small country enjoying a high standard of living may continue to depend for the purchase of industrial products on the export earnings of its primary products.

BASIC NEEDS OF INDUSTRY

The essential factors in the establishment of industry are capital, experienced management and technical skills, and markets. The raw materials used may either be indigenous or imported.

In an underdeveloped country the opportunities for the accumulation of capital out of internal savings will be poor and the process too slow for the needs of a rapid development programme. To meet the capital requirements of such a programme, two courses are open:

- (a) to encourage the investment of capital by established overseas concerns which may be willing to set up factories to exploit the local market potentialities
- (b) to negotiate foreign loans from international organizations, government agencies, or private sources.

---

* BAUER P. T. and YAMEY B. S. *The economics of under-developed countries.* Nisbet and Cambridge University Press, 1957.
† Under the term services it is usual to include, among others, trade, finance, government, transport, entertainment, catering and the public utilities. The goods-producing industries include agriculture, mining, manufacturing and construction.

146

Should industrial development have already made some headway, or should the country concerned have accumulated reserves of foreign exchange by the sale of primary products, as was not infrequently the case during the war, it may be possible to obtain at least a part of the necessary capital from internal resources. Current output of primary products, for the home market and for export, may be an important continuing source.

The provision of experienced management and the technical skills necessary for efficient production often presents a problem which can be solved only by the employment of expatriates recruited from those industrialized countries where such experience and skill reside. As an alternative to the direct employment of personnel, agreements may be made for the operation of a factory for a period of years by an established overseas concern, which may or may not have provided part of the capital. To be effective such agreements should cover a period sufficiently long to ensure that the experience necessary for efficient management can be imparted to those who will ultimately assume responsibility. Much help of this kind has already been provided under the Colombo Plan, the signatory countries of which afford training facilities for selected nationals of those countries whose own technological resources are inadequate for their needs.

Such measures and expedients are essentially transitional and as industry becomes more complex the adequacy of training facilities for technologists and skilled labour assumes increasing importance, and it becomes necessary to set up technical schools as part of the general educational system.

During the earlier phases of industrialization, production may be aimed at the internal market in satisfaction of the need for simple consumer goods which can be purchased with the cash arising from the sale of agricultural surpluses. Later, if the size of the country justifies it, industries could be established for the production of capital goods, machinery, etc., to provide the basis for further industrial expansion. The possibility of production for export should not be ignored, as the striking example of Hong Kong shows.

FOREIGN AID

This may take the form of providing special skills, money or goods. The special skills may be provided by advisers, managers, executives or technicians—whose periods of employment may be for days, weeks or years.

147

Grants and loans may take the form of goods, varying from such supplies as simple farm implements or tractors, to the equipment of a complete steel plant.

Loans may be of short or long duration. They may be tied to the purchase of goods from the lender's country or they may be freely convertible. They may be repayable in the borrower's own currency (soft) or in convertible currency (hard). Interest may be chargeable at specially low rates or the current market rate may be applied. Interest and repayment of the loan may commence immediately or it may be deferred for a number of years. In some cases loans offered at favourable rates may be associated with repayment in the lender's own currency.

Sources of aid include governments, international agencies, private foundations, banks and other commercial concerns, or the public at large when loans are issued for subscription.

When the aid consists of providing personnel, it has become usual for the agency concerned to pay the basic salary of the individual, together with his travelling expenses between his home and the country assisted. The borrower, however, is expected to meet the cost of the individual's local subsistence and of travel within the country to which he is assigned.

Such an arrangement means that the lender undertakes to meet the foreign exchange costs while the borrower is left to meet all costs incurred in local currency. This division of financial responsibility is intended to ensure that the borrower has a direct interest in seeing that the aid provided is used effectively.

Even when an agency undertakes to finance completely the construction of a project, the subsequent operating costs—which may be heavy—will have to be met by the borrower, since governments receiving economic aid are usually required to carry on the undertaking after the assistance given by the agency ceases. Thus, the acceptance of foreign aid may, if its nature is ill-judged, be expensive to the recipient and impose a heavy future drain on his internal resources.

It has become common since the war to tie foreign loans to a specific development project, a practice which has the aim of making a borrowing government act more responsibly. Such a definite tie between loan and project is, however, not essential provided there is reasonable political stability in the borrowing country. In the nineteenth century it was usual for governments to borrow upon the general security of their revenues without necessarily tying the loan to a specific project.

Provided the revenues of the borrowing country are buoyant, and it is considered likely to remain politically stable, there is no need to restrict the provision of foreign money to a self-liquidating project; indeed, it may often be the case that the provision of a loan to finance such works as roads and land reclamation will bring indirect benefits which far outweigh their direct cost.

It is clear that a foreign loan which is tied to purchases from the lending country weakens the negotiating position of the borrower when it comes to agreeing prices. This is especially so if the loan is tied to a particular manufacturer or group.

Whether a would-be borrower is able to obtain a loan for a specific development project depends in the first place on the confidence of the lender in the people to whom the loan is to be made, that is to say upon the lender's assessment of their credit-worthiness; such an assessment will take into account the past attitude adopted by the borrower towards foreign debts. In the second place it will depend upon the economic and technical soundness of the project concerned and upon the political stability of the country.

A final decision on whether a loan should be made will usually lie with professional administrators or financiers. To reach a decision they must rely upon the advice of technical experts possessing the necessary experience. Such experts may either form part of the lender's organization or they may be independent experts engaged on an *ad hoc* basis to examine all aspects of the project.

Of all the agencies to which a borrower may turn for assistance, by far the most important is the International Bank for Reconstruction and Development (The World Bank) which was set up in 1946 to meet the needs of a war-ravaged world. At first its activities were directed to providing loans to devastated European states to finance the importation of essential capital goods and raw materials, but in recent years it has

149

turned its attention increasingly to the economic development of the less advanced parts of the world.

Because its activities are of growing importance to engineers, a short account of the World Bank, its objectives and methods of operation has been included in Appendix D.

In concluding these remarks on the part played by the aid given by one country to another, it is worth remembering that nearly every industrialized country today has, in the past, benefited from foreign lending. For example, Dutch investors played a considerable part in financing British trade in the eighteenth century. Britain subsequently lent to almost every developing country in the world, including the United States. The United States borrowed heavily during the nineteenth century but has now developed into the world's greatest lender.

It should however be pointed out that in earlier days foreign lending was not regarded as a form of charitable assistance by one country to another, but as a financial bargain conveying mutual benefits to lender and borrower. There is today a school of thought which, while accepting the need in specified cases for direct assistance in the form of outright grants or loans at non-commercial rates of interest, considers that in general poorer countries will be best helped by expanding international trade and international lending on strictly commercial terms. This, it is argued, is less likely to lead to waste.

FEASIBILITY STUDIES

The first concern of a lending agency such as the World Bank is to assess the general economic situation of the borrower, and determine those sections of the economy to which priority should, in its view, be given. Some lenders would however take the view that their concern was only with the credit-worthiness of the borrower, or with the merits of the particular project for which finance was required, and not with more general economic appraisals.

Such a general appraisal would be followed by the selection of specific projects for closer study. This work would normally be undertaken by technical and economic experts and on their conclusions an opinion would be based as to the soundness of the proposed project.

150

The technical examination of any project should be carried out by teams possessing not only the necessary expert knowledge but a wide understanding of the social and economic problems of the country concerned. Since a general lack of accurate information and statistics is characteristic of underdeveloped countries, good judgement and experience in the evaluation of data assumes great importance.

It is important that the feasibility study should, so far as possible, present the estimated costs and benefits of a project in such a way as to permit a valid comparison with other possible alternative uses of the available funds. Such matters as the expected effect of the project on employment and other sociological aspects will sometimes be considered. It must however be emphasized that in the present state of knowledge conclusions reached on such matters are often based largely on personal judgement.

Allied with an examination of these aspects will be a review of the means of executing and subsequently managing the selected project. If the available organization is not considered adequate, the lender may well require the introduction of administrative changes and the employment of experienced agencies to ensure that any money made available will be spent efficiently.

INFRASTRUCTURE

An essential part of any general development plan is the provision made for public services and basic utilities—the infrastructure. These include transportation, communications, ports, flood control, land reclamation, irrigation, water supply, power, etc.

Investment in these services provides the foundation for the expansion of agriculture and manufacturing industry. The indirect benefits from such investment may, as has been pointed out, greatly exceed the direct revenues obtained. The direct revenues in some cases, such as those derived from roads and flood control works, may be nil, the costs involved being paid out of loans, general taxation, or grants from overseas government agencies.

The use of resources to provide such an infrastructure is characterized by a considerable lag between investment and its fruition. This lag is to a

11

large extent unavoidable since the works usually involve heavy expenditure spread over a number of years and are aimed at creating a service for which a use will only gradually develop. But while the provision of services in advance of their use represents an absorption of limited resources, it can seldom be entirely eliminated. The skill of the planner and the engineer lies in achieving the closest practicable match between availability and demand. Thus, the timing of a project is one of the keys to the economical use of resources.

Side by side with the expansion of basic services, provision should be made for government research into such matters as agricultural and mineral resources, water conservation, land drainage and reclamation.

SOPHISTICATION

Where new industries are concerned, the degree of sophistication which should be adopted calls for judgement and experience.

There is a well-known connexion between the wages paid and the capital employed in industry; those industries using the most capital-intensive methods of production tend to be associated with the payment of high per capita wages. The same is true of most nations: those most highly industrialized and using the greatest amount of energy per worker can be seen to enjoy a high standard of living.

Thus it is not unnatural that there should be a tendency to favour the adoption of the most modern methods of industrial production; a tendency which is fostered by the popular appeal of following closely in the van of technological progress and the desire for national prestige.

However, the premature adoption of processes that are too complex and too capital-intensive may create difficulties if the necessary managerial and technical skills have to be provided from abroad. And at a time when the problem is to provide useful employment for surplus labour it may be better to select techniques that will make a better use of easily-trained skills.

To this generalization there may be exceptions in which the adoption of advanced techniques may be justified on the grounds that the product is needed (for example, as part of the infrastructure) and that a small number of highly trained technicians is easier to recruit than the larger number of skilled craftsmen required by less advanced methods.

In conditions where there is a surplus of unskilled labour, the social cost to the community of employing it in industry or on public works may be negligible, although the direct money cost may be substantial.

In India, for instance, where the problem of unemployment is particularly acute, a number of large river development schemes have in recent years been carried out as a matter of policy by traditional methods employing the maximum of labour and the minimum of modern equipment. By adopting such a policy, useful work has been obtained from those whose productivity was previously virtually nil and scarce foreign exchange has been conserved for other purposes.

ENGINEERING STANDARDS

Allied with the selection of industrial processes is the choice of engineering standards appropriate to the local conditions.

The cost of providing a public service, whether it be a transportation system or a supply of electricity, may be significantly increased by trying to achieve too high a quality of service, since the additional costs incurred may be out of all proportion to the economic value of the improvement secured.

Even in highly industrialized countries, a tendency may be discerned to employ resources wastefully in an attempt to achieve 100% reliability for a public service. In an economic evaluation, the additional cost of providing a higher quality service should always be related to the estimated additional benefits that will ensue.

Standards can always be improved as time goes on and an economic need for a better service develops. A causeway carrying a road across a wide river, and liable to interrupt traffic during the flood season, may eventually be replaced by an all-weather bridge when the importance of the road to the regional economy makes the expenditure justified.

153

Again, in giving a supply of electricity to a sparse rural community it may be better, by lowering the technical specifications of the distribution system, to use the available capital resources to give, at a lower cost, a service which though not perfect is still acceptable to those who have to pay for it.

Appendix D   The World Bank

The International Bank for Reconstruction and Development (the World Bank) is a 'specialized agency' within the meaning of Article 57 of the Charter of the United Nations, and is required to function as an 'independent international organization'.

The Bank maintains working relations with other specialized international organizations including the International Monetary Fund (IMF), the Food and Agriculture Organization of the United Nations (FAO), the World Health Organization (WHO), the United Nations Educational, Scientific and Cultural Organization (UNESCO), and the United Nations Development Programme (UNDP).

The Bank was founded, along with its sister organization, the IMF, at the International Financial and Monetary Conference at Bretton Woods, New Hampshire, USA, in July, 1944.

The two institutions have complementary functions. That of the IMF is to promote international currency stability, while that of the Bank is to help create a more prosperous and better-balanced world economy.

The first loans made by the Bank were to European countries to assist them with post-war reconstruction, but in 1948 the Bank turned its attention to its other major responsibility—the promotion of economic development in its poorer member countries. This has remained its chief preoccupation.

The capital of the Bank is subscribed by member governments, which now number 109, in accordance with their economic strength. Only about one tenth of the subscribed capital ($22 900 million in June 1968) has been paid-up, the remainder being at call if required by the Bank to meet its obligations on borrowings or on loans guaranteed by it. Borrowings in world capital markets are the main source of the Bank's resources, the guarantee afforded by the Bank's capital structure making it possible to mobilize private capital through bond issues.

The Bank's Charter contains a number of protective provisions governing loans or guarantees made by the Bank. Among these provisions are that:
  (a) the loans must be for productive purposes
  (b) except in special circumstances, the loans must be used to finance the foreign exchange requirements of specific projects
  (c) the merits of all projects must be carefully examined and priority given to the most urgent
  (d) the borrower may be a member government or a non-governmental enterprise; but if the latter, the loan must be backed by a government guarantee

Under the Articles of Agreement, the Bank is also required to exercise 'prudence' in making loans and to consider carefully the ability of the borrower to meet its repayment obligations. In addition, the Bank is required to ensure that the loan is used solely for the specific purpose for which it was granted.

No conditions are attached to the Bank's loans requiring their proceeds to be spent in the territory of any particular member or members.

The Bank requires its borrowers to obtain goods and services purchased with Bank finance through international competitive bidding unless this procedure is clearly inappropriate. In some cases borrowers are requested to employ the services of consultants to assist in determining the qualifications of bidders and in analysing bids.

The granting of a loan is contingent upon the Bank being satisfied that the borrower is unable to obtain a loan on acceptable terms from other sources.

A part of the Bank's activities which has assumed increasing importance in recent years is the provision of technical assistance, either by the Bank's own staff or by independent consultants.

Most developing countries need assistance in identifying and preparing projects for financing. They may also need help with formulating appropriate development policies, establishing effective development institutions, determining priorities for investment, and other tasks essential to their development. The Bank has been requested, or found it necessary, to advise on all these and other matters at some time or another over the years.

Another sphere in which the Bank has played a part has been in helping to settle disputes between its member countries. Two notable examples are the settlement, in 1960, of the dispute over the sharing of the Indus Waters between India and Pakistan, and the settlement of compensation claims arising out of the nationalization of the Suez Canal in 1956.

It has been stated that under its Articles, the Bank is required to examine carefully the ability of the borrower to meet its repayment obligations. Service payments on Bank loans are in foreign currency. Before a loan is provided, an appraisal is made of whether the amount contemplated is within the limits which the prospective borrowing country can reasonably be expected to service, taking account not only of its existing and prospective debts to the Bank, but of all sources of external finance.

In the case of a country whose credit is impaired by the existence of a dispute over a default on its foreign debt or over compensation for expropriated property formerly owned by foreigners, the Bank's normal practice is to inform the government involved that the Bank will not assist it unless and until it makes appropriate efforts to reach a fair and equitable settlement.

The Bank attaches the greatest importance to assuring itself that a particular project for which a loan is required is not only economically and technically sound, but that its order of priority in relation to other possible projects has been carefully determined in the light of the overall development requirements of the country concerned. The Bank will not make loans to cover vague or unspecified development programmes.

Normally a Bank loan is related to the expenditure on imported goods and services required in carrying out a project. The remainder of the financing must normally be provided or found by the borrower.

The rate of interest charged by the Bank on its loans is based on the cost to the Bank of raising money in the world's capital markets. It is kept as low as is compatible with the need to maintain the Bank's financial strength and reputation.

As a result of the Bank's experience in providing loans to underdeveloped countries, the International Development Association (IDA) was formed in 1960 to meet the needs of countries which are too poor to service 'hard' loans at normal rates of interest. For a nominal charge of 0·75% IDA provide interest-free loans covering terms up to 50 years and allowing up to 10 years before the repayments commence.

The resources of IDA are provided by subscriptions from the industrial countries and by transfer from the funds of the Bank.

Another form of assistance which has grown out of the Bank's activities is that provided by the International Finance Corporation (IFC), the purpose of which is to promote industrial development by private enterprise without a government guarantee.

Appendix E   Amortization

Every facility, whether it be an undertaking devoted to electricity supply, to transport, or the provision of some other service or product, is subject to economic change as the result of physical deterioration or of obsolescence arising from the march of knowledge, and the changing tastes and desires of people. As a consequence, physical assets represented by an investment may one day be unable to serve economically the purpose for which they were originally created or may no longer be needed for that purpose.

This is recognized in the economic evaluations discussed in this handbook. In these evaluations it is necessary to estimate the period during which the purchasing power originally invested in the asset must be recovered in cash or justified by non-cash benefits. This period is sometimes known as the amortization period, and the process of recovery as amortization.

This fact is similarly recognized in conventional accounting measurements of income or profit, in which the estimated value that an asset will lose over its life is set off against revenue earned. As accounting reports are usually made annually, a formula is used to allot to each year its proportion of the total loss in value. The resulting figure is usually called a provision for depreciation, though strictly it is a measurement rather than a provision.

PHYSICAL DETERIORATION

From the moment that a piece of machinery or a structure is put into service it begins to wear out, imperceptibly at first, more noticeably as time goes on. By spending money on the replacement of worn parts or by making good physical defects the effect of such deterioration can be kept in check and, theoretically at least, the process of gradual replacement or repair can be continued indefinitely. In practice, however, a point is

159

eventually reached at which it no longer becomes profitable to spend money on keeping an old machine or structure in repair; it may then be said to have reached the end of its useful life and to require replacement as the result of general physical deterioration.

OBSOLESCENCE

Obsolescence may be defined as the loss of economic value as the result of change in demand (as when tastes in clothing change) or of the development, due to scientific progress, of equipment or methods which provide a similar service more efficiently or at a lower cost.

Many examples of obsolescence arising from technological change or changes in demand conditions can be given. They include the following.

Surface transport

The later part of the eighteenth and the beginning of the nineteenth centuries saw the heyday of canal transportation in Great Britain, which may be said to have reached its culmination with the opening of the Caledonian Canal in 1822. By that time, however, the economic background which had seemed so bright when the project was sanctioned in 1803 had already changed. Steam propulsion had been introduced for coast-wise shipping and had made vessels independent of the winds, while the development of the steam locomotive for land transport was already looming as a serious threat to the future of canal transportation. Thereafter, the demand for the services of the canals steadily declined as their traffics were diverted to the railways, and one after another they fell into increasing disuse.

A century later the railways of Great Britain were in their turn to feel the economic effect of the changed conditions which had arisen from the development of fast road transport, made possible by the introduction of the internal-combustion engine. In the period between the two world wars, many rural lines were taken out of service under pressure from road competition, and this process is today continuing at an increasing tempo.

Thermal-electric power

In the field of electric-power generation the advent of the steam turbine effected a revolution in the design of power-station equipment, and during a few short years at the beginning of the present century brought about a position in which it was economically sound to pull out and scrap, in favour of the more efficient turbine, steam reciprocating engines which had been installed only a few years previously. But apart from technical developments, changes in the level of demand may operate to render the equipment of a power undertaking obsolete.* The gradual increase in the size of the load to be served may make it economical to replace cables or switchgear while they are still physically capable of giving many years of satisfactory service. The decline of local industries or the movement of population may reduce the earning capacity of a particular feeder or distribution network and so render the equipment redundant and unprofitable. Or the necessity to increase greatly the generating capacity in a particular area may lead to the abandonment of a power station on a restricted site which does not permit of large-scale extension. Changes may come suddenly as the result of the introduction of some fundamentally new principle into the science of power production, or they may creep in slowly and almost unobserved.

Perhaps the most profound effect of technological change on the development of the power industry has been provided by nuclear physics. The first nuclear power station in the world was inaugurated at Calder Hall in Cumberland in October 1956, an event which was preceded the year before by the decision of the British Government to press ahead with a programme of nuclear power station construction. It was not anticipated that the power stations to be built under this programme would be competitive with conventional steam stations using coal or oil fuel, but it was held at the time that the immense capital investment envisaged was justified on the grounds that it represented the best way of closing the anticipated energy gap.

Since 1956 succeeding nuclear power stations have demonstrated the economic advantages to be derived both from steady progress in technological development and from increasing size. As a consequence a gradual reduction in the price gap between power stations using fossil fuels and those using nuclear fuels has been achieved. The adoption by the British electricity authorities in 1966 of the Advanced Gas-cooled

---

* Note that obsolescence is related to a particular need. The equipment may still have a saleable value, say to an overseas power company with a lower level of demand; and it may be retained for stand-by use.

Reactor as exemplified by the Dungeness B Power Station, made it possible to forecast that early in the 1970's nuclear power will be produced at a cost which will show a significant saving over that of the most economical steam stations using fossil fuels which will then be coming into use. Indeed, it was possible in 1967 for the Central Electricity Generating Board to claim that the later magnox stations such as Sizewell, located in areas remote from the cheaper sources of coal, were already producing electricity at costs comparable with those of coal-fired stations in similar areas.

Because the economic aspects of electric power generation are not only complex but continually changing, the risk that a technical decision taken as the result of a competent assessment of the prevailing situation may, within a few years, be open to criticism, cannot be ignored.

In a period of less rapid change between the two wars, British electricity authorities assumed, for accounting purposes, that all power station equipment should be written-off within a period of 20 years. Subsequent experience, however, demonstrated that this assumption was unnecessarily pessimistic and that an effective life of conventional steam plant of 30 years or even more was by no means uncommon.

In the nuclear field, although it is too early to predict the working lives of power stations constructed under the 1955 programme, experience to date nevertheless suggests that they are likely to remain effective for periods considerably in excess of the 20 years which was originally assumed.

As in any economic comparison, that between alternative ways of producing electricity should be based upon the use of equally realistic lives for the principal components concerned. Should the amortization periods not be equally valid, any comparison based upon them will be distorted and may well lead to erroneous decisions.

Water power
The component parts of a hydroelectric undertaking are less susceptible both to physical deterioration and to obsolescence than is the case with a thermal station. The bulk of the assets of hydro undertakings consists of heavy civil engineering works such as dams and tunnels which can, with proper maintenance, be kept in a state of almost perpetual usefulness.

Again, the operating costs form but a fraction of the total cost of producing energy, the greater part of which is represented by the fixed interest and other charges associated with the original capital investment. Moreover, since the overall efficiency of the mechanical and electrical equipment used leaves little scope for future improvement, there is no incentive to replace it before it becomes physically worn out.

In spite of these advantages, however, many of the early hydro power plants have not escaped the effect of obsolescence as the result of changing conditions. The great majority of these early schemes were the product of private enterprise and were established either to meet the needs of a local community or to support a specially created industry. Since, for commercial reasons, it was essential on the one hand to produce power at the lowest practicable cost and on the other to avoid any very large commitment of capital during the early years of the undertaking, sites were chosen which permitted the rapid creation of a source of low-cost power without regard to the ultimate utilization of the regional resources. A consequence of this policy was that some of the old schemes made but a partial use of the potentialities of the site. This has led in certain cases to the abandonment of the original development and the incorporation of the site in a much larger scheme designed to make a full use of the area resources. This does not imply that the early schemes were necessarily unsound.

Premature abandonment

An outstanding example of the early retirement of an important 'permanent' structure is provided by the Dixence Dam in Switzerland, completed in 1935. This dam which has a height of 85 m was, at the time it was built, amongst the highest in Europe, yet within 13 years of its construction the decision was taken to build a very much larger dam immediately downstream of the present structure. The new dam, completed in 1964, has a maximum height of 276 m, and creates a reservoir which completely submerges the old dam which has thus become useless.

The early abandonment of the original Dixence Dam was held to be justified by the benefits to be secured from the redevelopment of this high Alpine valley as a greatly enlarged reservoir. It represented a more intensive development of the water power resources of the region to meet the demand for winter electrical energy which had been increasing at a

163

rate which could hardly have been foreseen when the original dam was built. The justification for such a development must be tested by carrying out an economic appraisal on the lines described in this handbook.

Effects of rapid change

It is not only the technology of electrical energy production which has been subject to increasingly rapid change but also that of almost every field of engineering endeavour. The problem to be answered, in selecting an appropriate amortization period, is only rarely one of deciding when a piece of equipment or a structure will wear out but nearly always how long is it likely to serve efficiently its original purpose. The increasing speed of change makes it especially difficult to estimate the effective life which should be assigned to investments in the energy field, and the history of the post-war years is littered with forecasts carefully made on the basis of the best information available at the time, but which were falsified by events within a few years.

The gas industry affords a good example of the effect of unforeseen technical advance. For many years it had been declining in importance as a form of energy and it seemed doomed eventually to die out. Since the late 1950s, however, the situation has changed radically and it has become hazardous to make a forecast of the impact which the gas industry is likely to have on the British economy over the next few years. The balance of advantages as between different kinds of fuel is continually liable to change as the result of events which are neither predictable nor controllable.

But in spite of the difficulties that beset those engineering activities which are subject to rapid change there are certain fields, such as those associated with water resource development, whether for potable purposes, for power, or for irrigation, which are at present less prone to the effects of change than are those concerned with energy or industrial processes. The works concerned are usually fairly simple and robust in character and provide for human needs by harnessing natural resources, the ultimate extent of which can be determined in advance with some confidence while the pattern of future use is less susceptible to change. In these cases it may be legitimate to assume that the effective lives of the principal works are closely allied to their estimated physical lives—due regard being paid to the possibility that foreseeable changes may require the replacement of equipment which is still physically serviceable.

EFFECTIVE LIVES

For purposes of economic evaluation and of accounting it has been common to provide for the total amortization of heavy civil engineering works over periods usually ranging from 50 to 100 years. In the case of dams, the period chosen has sometimes been determined by the terms under which the water rights were acquired when the dam was built, or more rarely, by the estimated life of the reservoir where this is likely to be seriously limited by siltation. Where a 'permanent' structure is involved, selection of an appropriate amortization period depends very much on judgement, taking into account an assessment of its probable physical life, modified by the current view of the obsolescence factor and by the general policies of the owners.

Table E.1 gives a range of lives often assumed for the purpose of providing for the depreciation of physical assets.

TABLE E.1

	Range in years	
	From	To
Heavy civil engineering works including dams, tunnels, canals, docks, harbours and breakwaters	50	100
Permanent buildings and roads, including bridges, in steel, stone, brick or concrete	40	60
Outdoor steel structures including flood and lock gates, cranes and steel pipelines	35	50
Underground electric cables	35	40
Overhead electric power lines on steel towers, water turbines, electric generators and associated equipment	30	35
Plant and equipment for thermal power stations and sub-stations, including boilers, turbo-alternators, transformers and switchgear	25	30
Railway tracks	20	35
Tugs, dredgers and floating cranes	20	30
Heating and ventilating equipment, lifts	20	25
Diesel-electric stations	15	20
Portable tools and equipment, office furniture	12	15
Road vehicles	5	8

It is usual to make no allowance for the depreciation of freehold land on the ground that the land itself is indestructible and is likely, in the long run, to increase in value. In the case of leasehold land and buildings, mining, water rights and other concessions, the amortization period chosen should not exceed the unexpired portion of the lease or concession. Buildings provided expressly to house particular equipment should not normally be assigned a longer life than that of the equipment itself.

The assets covered by the above table fall into two classes:

    (a) heavy civil engineering works which can be kept in good condition over a long term of years

    (b) steel structures, mechanical and electrical equipment, etc., which are exposed to both physical deterioration and, in some cases, to risks of rapid obsolescence.

Choice of a suitable amortization period for the first class of asset is somewhat arbitrary, depending on the policy adopted by the owning authority.* The longer lives given in Table E.1 may be regarded as a reasonable estimate of the physical life of assets which have been well constructed but they take no account of the possibility that their usefulness may be curtailed as the result of obsolescence. Amortization periods for the second class of assets depend partly upon experience of actual physical deterioration and partly on the views held on obsolescence.

In most cases the cost of providing for the replacement of long-life assets is not likely to be significant in comparison with the interest charge especially where the interest rates are high. This is demonstrated in Table E.2.

TABLE E.2

Amortization period	Equivalent annuity as a percentage of initial capital value at interest rate:		
	4%	8%	12%
50 years	4·655%	8·174%	12·0004%
100 years	4·081%	8·004%	12·0000%
Infinite life	4·000%	8·000%	12·0000%

* For instance it is the practice of both the US Bureau of Reclamation and the US Corps of Engineers to make provision for the amortization of the whole of their assets within a period of 50 years. If the useful life of individual assets, such as electrical and mechanical equipment, is expected to be less than 50 years, the estimates for the operation and maintenance of the projects include provision for necessary replacements. The method of calculation used has the effect of not taking credit for the residual value of major replacements which would have a remaining useful physical life at the expiration of the 50 year amortization period assumed for the project as a whole.

The annual capital charges for assets with lives of 50 or 100 years differ little from the annual charge for an asset with an infinite life, which consists simply of the interest charge with no element for recovery of the cost of the asset. Expressed in terms of the total equivalent annual cost of the works the additional burden of reducing the amortization period from 100 years to 50 years is not likely to affect significantly the economic presentation of a project.

PROVISION FOR DEPRECIATION IN CONVENTIONAL ACCOUNTING REPORTS

This is usually made by taking the original cost of the asset concerned, including all engineering expenses, and spreading the sum uniformly over the life assigned to it. This is the straight line method of depreciation. It is evident that in times of inflation the sum of such annual amounts will be less than the value of the original asset when this is corrected for the rise in the price level. Similarly, in each year the financial report will understate the annual depreciation provision as expressed in current price level terms. Since such accounting data may be used among other things as appropriate guides in pricing, this is a matter for concern.

DEPRECIATION ON A REPLACEMENT COST BASIS

To meet such a situation many authorities have suggested that the annual depreciation provision should be based on the estimated replacement cost of the asset in question rather than on the historic cost. This can only provide a rough guide since it is highly unlikely that any replacement will be an exact duplicate of the original asset, but it has been adopted in some cases. It may be noted that the two Government White Papers dealing with the economic and financial objectives of the nationalized industries (Command 1337 and 3437) require them to show in their accounts, as a deduction from their revenues, the annual depreciation of their assets on an historic cost basis, and in addition to report such further amounts as may be necessary to make up the difference between historic and replacement costs.

The Revenue Account for the year ended 31 December, 1967, of the British Transport Docks Board includes under 'Depreciation' the sum of £367 000 as 'Additional amount to reflect changes in the purchasing power of money'.

In their 'Notes on the accounts' the Docks Board explain that their fixed assets would have been valued at some £104 million on the basis of their cost to the Board but that these have been 'written up' to a total of £118 million, representing the estimated value of the assets in terms of current purchasing power of money. The increase in this case has been calculated from the average index of the purchasing power of money in 1967 as compared with the corresponding index for January 1963 (the vesting date of the Docks Board). It can be argued that this is a more appropriate basis for such an adjustment, as what is being recorded is essentially the loss of general purchasing power that was originally invested in the asset.

Not only the nationalized industries but many commercial companies now make increased provision in their accounts to cover the renewal of their assets on a replacement cost basis.

Appendix F   Compound interest and annuity tables

The following tables contain values for interest formulae which are sufficient for most discounted cash flow and interest calculations that an engineer will need.   For extensive work more comprehensive tables such as those published by Oliver and Boyd or Shaw and Sons may be found necessary.

These tables follow the customary practice of assuming that all payments are made at the end of the year and that interest is compounded annually. It is possible to adjust the values to make them correct for more frequent intervals of compounding.   In many applications, such refinement makes no significant difference to the results.

NOTES ON COMPOUND INTEREST AND ANNUITY TABLES

A.   *Terminal value of a single sum at compound interest* (see Table 1)
The amount S to which £1 will increase in n years with interest r.
At end of year: 1 $S = 1+r$
$\quad\quad\quad\quad\quad$ 2 $S = 1+r+r(1+r) = (1+r)^2$
$\quad\quad\quad\quad\quad$ 3 $S = (1+r)^2+r(1+r)^2 = (1+r)^3$
$\quad\quad\quad\quad\quad$ n $S = (1+r)^{n-1}+r(1+r)^{n-1} = (1+r)^n$
Thus $S=(1+r)^n$

B.   *Present value of a single sum* (see Table 2)
The present value A, of £1 n years hence, when discounted at interest r.
The present value of S, n years hence, in note A, is £1.   Consequently the present value A of £1 n years hence is

$$A = \frac{1}{S} = \frac{1}{(1+r)^n} = (1+r)^{-n}$$

C.   *Present value of an annuity* (see Table 3)
The present value A of £1 per annum for n years when discounted at interest r.

For a series of: 1 payment $\quad A = (1+r)^{-1}$

$\qquad\qquad\qquad$ 2 payments $\quad A = (1+r)^{-1}+(1+r)^{-2}$

$\qquad\qquad\qquad$ n payments $\quad A = (1+r)^{-1}+(1+r)^{-2}\ldots+(1+r)^{-n}$

A is then a geometric progression with n terms which can be summed by subtracting from it $A(1+r)$.

$$A(1+r) = 1+(1+r)^{-1}\ldots+(1+r)^{-n+1}$$
$$A-A(1+r) = -1+(1+r)^{-n}$$

hence $\qquad Ar = 1-(1+r)^{-n}$

$$A = \frac{1-(1+r)^{-n}}{r}$$

The terminal value S at the end of year n of A accumulated at rate r will be $A(1+r)^n$ (see note A)

whence

$$S = \frac{(1+r)^n-1}{r}$$

This expression is the same as that for the terminal value of an annuity accumulated for n years at rate r (see note E).

Similarly the present value A of a series of payments each of £S every n years can be expressed as

$$A = S\left\{\frac{1}{(1+r)^n}+\frac{1}{(1+r)^{2n}}+\ldots\right\}$$

hence $\quad A(1+r)^n = S\left\{1+\frac{1}{(1+r)^n}+\frac{1}{(1+r)^{2n}}+\ldots\right\}$

Subtracting the first equation from the second equation

$$A\{(1+r)^n-1\} = S$$

or

$$A = \frac{S}{(1+r)^n-1}$$

which on conversion for easy use of tables is

$$S\left\{\frac{r}{(1+r)^n-1}\right\}\left(\frac{1}{r}\right)$$

D. *Redemption of a loan by equal annual payments* (see Table 3)

The amount P per annum to redeem a loan of £1 (or to recover a sum of £1) over n years at interest r on the outstanding balance.

The loan, initially £1, is reduced each year by the portion of the annual amount P left over from payment of interest on the outstanding debt.

The outstanding debt becomes at end of

year: 1 $1+r-P$

2   $(1+r-P)(1+r)-P = (1+r)^2-P(1+r)-P$

3   $\{(1+r)^2-P(1+r)-P\}(1+r)-P = (1+r)^3$

     $-P\{(1+r)^2+(1+r)+1\}$

n $(1+r)^n-P\{(1+r)^{n-1}+(1+r)^{n-2}\ldots+1\}$

which by definition is zero, whence

$$(1+r)^n = P\{(1+r)^{n-1}+(1+r)^{n-2}\ldots+1\}$$

and $1 = P\{(1+r)^{-1}+(1+r)^{-2}\ldots(1+r)^{-n}\} = PA$

(A being the geometric progression given in note C)

whence $P = \dfrac{1}{A} = \dfrac{r}{1-(1+r)^{-n}}$

It is not surprising that this expression is the reciprocal of that for the present value of an annuity of £1 since the amount loaned can be visualized as equal to the total of the present values of its future redemption payments, that is, it provides an annuity for the lender.

E.   *Sinking fund* (see Table 4)

The amount P per annum for n years at interest rate r needed to accumulate to £1.

With an annual payment the accumulated amount S becomes at end of

year: 1 P

2   $P(1+r)+P$

n $P\{(1+r)^{n-1}+(1+r)^{n-2}\ldots+1\}$

Again S is a geometric progression which can be summed by subtracting from it $S(1+r)$ to give

$$Sr = P\{(1+r)^n-1\}$$

or $S = P\left\{\dfrac{(1+r)^n-1}{r}\right\}$

S by definition is £1, whence

$$P = \frac{r}{(1+r)^n-1}$$

If now $P=$£1,

$$S = \frac{(1+r)^n-1}{r} \quad \text{(see note C)}$$

TABLE 1. TERMINAL VALUE OF A SINGLE SUM AT COMPOUND INTEREST

The amount to which £1 will increase in n years with interest rate r per annum $= (1+r)^n$. (See note A)

Interest % ($=100r$)

n (years)	1·0	1·5	2·0	2·5	3·0	3·5	4·0	4·5	5·0	5·5
1	1·0100	1·0150	1·0200	1·0250	1·0300	1·0350	1·0400	1·0450	1·0500	1·0550
2	1·0201	1·0302	1·0404	1·0506	1·0609	1·0712	1·0816	1·0920	1·1025	1·1130
3	1·0303	1·0457	1·0612	1·0769	1·0927	1·1087	1·1249	1·1412	1·1576	1·1742
4	1·0406	1·0614	1·0824	1·1038	1·1255	1·1475	1·1699	1·1925	1·2155	1·2388
5	1·0510	1·0773	1·1041	1·1314	1·1593	1·1877	1·2167	1·2462	1·2763	1·3070
6	1·0615	1·0934	1·1262	1·1597	1·1941	1·2293	1·2653	1·3023	1·3401	1·3788
7	1·0721	1·1098	1·1487	1·1887	1·2299	1·2723	1·3159	1·3609	1·4071	1·4547
8	1·0829	1·1265	1·1717	1·2184	1·2668	1·3168	1·3686	1·4221	1·4775	1·5347
9	1·0937	1·1434	1·1951	1·2489	1·3048	1·3629	1·4233	1·4861	1·5513	1·6191
10	1·1046	1·1605	1·2190	1·2801	1·3439	1·4106	1·4802	1·5530	1·6289	1·7081
11	1·1157	1·1779	1·2434	1·3121	1·3842	1·4600	1·5395	1·6229	1·7103	1·8021
12	1·1268	1·1956	1·2682	1·3449	1·4258	1·5111	1·6010	1·6959	1·7959	1·9012
13	1·1381	1·2136	1·2936	1·3785	1·4685	1·5640	1·6651	1·7722	1·8856	2·0058
14	1·1495	1·2318	1·3195	1·4130	1·5126	1·6187	1·7317	1·8519	1·9799	2·1161
15	1·1610	1·2502	1·3459	1·4483	1·5580	1·6753	1·8009	1·9353	2·0789	2·2325
16	1·1726	1·2690	1·3728	1·4845	1·6047	1·7340	1·8730	2·0224	2·1829	2·3553
17	1·1843	1·2880	1·4002	1·5216	1·6528	1·7947	1·9479	2·1134	2·2920	2·4848
18	1·1961	1·3073	1·4282	1·5597	1·7024	1·8575	2·0258	2·2085	2·4066	2·6215
19	1·2081	1·3270	1·4568	1·5986	1·7535	1·9225	2·1068	2·3079	2·5269	2·7656
20	1·2202	1·3469	1·4859	1·6386	1·8061	1·9898	2·1911	2·4117	2·6533	2·9178
25	1·2824	1·4509	1·6406	1·8539	2·0938	2·3632	2·6658	3·0054	3·3864	3·8134
30	1·3478	1·5631	1·8114	2·0976	2·4273	2·8068	3·2434	3·7453	4·3219	4·9840
35	1·4166	1·6839	1·9999	2·3732	2·8139	3·3336	3·9461	4·6673	5·5160	6·5138
40	1·4889	1·8140	2·2080	2·6851	3·2620	3·9593	4·8010	5·8164	7·0400	8·5133
45	1·5648	1·9542	2·4379	3·0379	3·7816	4·7024	5·8412	7·2482	8·9850	11·127
50	1·6446	2·1052	2·6916	3·4371	4·3839	5·5849	7·1067	9·0326	11·467	14·542
55	1·7286	2·2679	2·9717	3·8888	5·0821	6·6331	8·6464	11·256	14·636	19·006
60	1·8167	2·4432	3·2810	4·3998	5·8916	7·8781	10·519	14·027	18·679	24·840

Interest % (=100r)

n (years)	6·0	6·5	7·0	7·5	8·0	9·0	10·0	12·0	15·0	20·0
1	1·0600	1·0650	1·0700	1·0750	1·0800	1·0900	1·1000	1·1200	1·1500	1·2000
2	1·1236	1·1342	1·1449	1·1556	1·1664	1·1881	1·2100	1·2544	1·3225	1·4400
3	1·1910	1·2079	1·2250	1·2423	1·2597	1·2950	1·3310	1·4049	1·5209	1·7280
4	1·2625	1·2865	1·3108	1·3355	1·3605	1·4116	1·4641	1·5735	1·7490	2·0736
5	1·3382	1·3701	1·4026	1·4356	1·4693	1·5386	1·6105	1·7623	2·0114	2·4883
6	1·4185	1·4591	1·5007	1·5433	1·5869	1·6771	1·7716	1·9738	2·3131	2·9860
7	1·5036	1·5540	1·6058	1·6590	1·7138	1·8280	1·9487	2·2107	2·6600	3·5832
8	1·5938	1·6550	1·7182	1·7835	1·8509	1·9926	2·1436	2·4760	3·0590	4·2998
9	1·6895	1·7626	1·8385	1·9172	1·9990	2·1719	2·3579	2·7731	3·5179	5·1598
10	1·7908	1·8771	1·9672	2·0610	2·1589	2·3674	2·5937	3·1058	4·0456	6·1917
11	1·8983	1·9992	2·1049	2·2156	2·3316	2·5804	2·8531	3·4785	4·6524	7·4301
12	2·0122	2·1291	2·2522	2·3818	2·5182	2·8127	3·1384	3·8960	5·3502	8·9161
13	2·1329	2·2675	2·4098	2·5604	2·7196	3·0658	3·4523	4·3635	6·1528	10·699
14	2·2609	2·4149	2·5785	2·7524	2·9372	3·3417	3·7975	4·8871	7·0757	12·839
15	2·3966	2·5718	2·7590	2·9589	3·1722	3·6425	4·1772	5·4736	8·1371	15·407
16	2·5404	2·7390	2·9522	3·1808	3·4259	3·9703	4·5950	6·1304	9·3576	18·488
17	2·6928	2·9170	3·1588	3·4194	3·7000	4·3276	5·0545	6·8660	10·761	22·186
18	2·8543	3·1067	3·3799	3·6758	3·9960	4·7171	5·5599	7·6900	12·375	26·623
19	3·0256	3·3086	3·6165	3·9515	4·3157	5·1417	6·1159	8·6128	14·232	31·948
20	3·2071	3·5236	3·8697	4·2479	4·6610	5·6044	6·7275	9·6463	16·367	38·338
25	4·2919	4·8271	5·4274	6·0983	6·8485	8·6231	10·835	17·000	32·919	95·396
30	5·7435	6·6144	7·6123	8·7550	10·063	13·268	17·449	29·960	66·212	237·38
35	7·6861	9·0623	10·677	12·569	14·785	20·414	28·102	52·800	133·18	590·67
40	10·286	12·416	14·974	18·044	21·725	31·409	45·259	93·051	267·86	1469·8
45	13·765	17·011	21·002	25·905	31·920	48·327	72·890	163·99	538·77	3657·3
50	18·420	23·307	29·457	37·190	46·902	74·358	117·39	289·00	1083·7	9100·4
55	24·650	31·932	41·315	53·391	68·914	114·41	189·06	509·32	2179·7	22644
60	32·988	43·750	57·946	76·649	101·26	176·03	304·50	897·59	4384·1	56346

173

TABLE 2. PRESENT VALUE OF A SINGLE SUM

The present value of £1 n years hence, when discounted at interest rate r per annum $= (1+r)^{-n}$. (See note B)

Interest % $(=100r)$

n (years)	1	1·5	2	2·5	3	3·5	4	4·5	5	5·5
1	0·99010	0·98522	0·98039	0·97561	0·97087	0·96618	0·96154	0·95694	0·95238	0·94787
2	0·98030	0·97066	0·96117	0·95181	0·94260	0·93351	0·92456	0·91573	0·90703	0·89845
3	0·97059	0·95632	0·94232	0·92860	0·91514	0·90194	0·88900	0·87630	0·86384	0·85161
4	0·96098	0·94218	0·92385	0·90595	0·88849	0·87144	0·85480	0·83856	0·82270	0·80722
5	0·95147	0·92826	0·90573	0·88385	0·86261	0·84197	0·82193	0·80245	0·78353	0·76513
6	0·94205	0·91454	0·88797	0·86230	0·83748	0·81350	0·79031	0·76790	0·74622	0·72525
7	0·93272	0·90103	0·87056	0·84127	0·81309	0·78599	0·75992	0·73483	0·71068	0·68744
8	0·92348	0·88771	0·85349	0·82075	0·78941	0·75941	0·73069	0·70319	0·67684	0·65160
9	0·91434	0·87459	0·83676	0·80073	0·76642	0·73373	0·70259	0·67290	0·64461	0·61763
10	0·90529	0·86167	0·82035	0·78120	0·74409	0·70892	0·67556	0·64393	0·61391	0·58543
11	0·89632	0·84893	0·80426	0·76214	0·72242	0·68495	0·64958	0·61620	0·58468	0·55491
12	0·88745	0·83639	0·78849	0·74356	0·70138	0·66178	0·62460	0·58966	0·55684	0·52598
13	0·87866	0·82403	0·77303	0·72542	0·68095	0·63940	0·60057	0·56427	0·53032	0·49856
14	0·86996	0·81185	0·75788	0·70773	0·66112	0·61778	0·57748	0·53997	0·50507	0·47257
15	0·86135	0·79985	0·74301	0·69047	0·64186	0·59689	0·55526	0·51672	0·48102	0·44793
16	0·85282	0·78803	0·72845	0·67363	0·62317	0·57671	0·53391	0·49447	0·45811	0·42458
17	0·84438	0·77637	0·71416	0·65720	0·60502	0·55720	0·51337	0·47318	0·43630	0·40245
18	0·83602	0·76491	0·70016	0·64117	0·58739	0·53836	0·49363	0·45280	0·41552	0·38147
19	0·82774	0·75361	0·68643	0·62553	0·57029	0·52016	0·47464	0·43330	0·39573	0·36158
20	0·81954	0·74247	0·67297	0·61027	0·55368	0·50257	0·45639	0·41464	0·37689	0·34273
25	0·77977	0·68921	0·60953	0·53939	0·47761	0·42315	0·37512	0·33273	0·29530	0·26223
30	0·74192	0·63976	0·55207	0·47674	0·41199	0·35628	0·30832	0·26700	0·23138	0·20064
35	0·70591	0·59387	0·50003	0·42137	0·35538	0·29998	0·25342	0·21425	0·18129	0·15352
40	0·67165	0·55126	0·45289	0·37243	0·30656	0·25257	0·20829	0·17193	0·14205	0·11746
45	0·63905	0·51171	0·41020	0·32917	0·26444	0·21266	0·17120	0·13796	0·11130	0·08988
50	0·60804	0·47500	0·37153	0·29094	0·22811	0·17905	0·14071	0·11071	0·08720	0·06877
55	0·57853	0·44093	0·33650	0·25715	0·19677	0·15076	0·11566	0·08884	0·06833	0·05262
60	0·55045	0·40930	0·30478	0·22728	0·16973	0·12693	0·09506	0·07129	0·05354	0·04026

Interest % (=100r)

n (years)	6	6·5	7	7·5	8	9	10	12	15	20
1	0·94340	0·93897	0·93458	0·93023	0·92593	0·91743	0·90909	0·89286	0·86957	0·83333
2	0·89000	0·88166	0·87344	0·86533	0·85734	0·84168	0·82645	0·79719	0·75614	0·69444
3	0·83962	0·82785	0·81630	0·80496	0·79383	0·77218	0·75131	0·71178	0·65752	0·57870
4	0·79209	0·77732	0·76290	0·74880	0·73503	0·70843	0·68301	0·63552	0·57175	0·48225
5	0·74726	0·72988	0·71299	0·69656	0·68058	0·64993	0·62092	0·56743	0·49718	0·40188
6	0·70496	0·68533	0·66634	0·64796	0·63017	0·59627	0·56447	0·50663	0·43233	0·33490
7	0·66506	0·64351	0·62275	0·60275	0·58349	0·54703	0·51316	0·45235	0·37594	0·27908
8	0·62741	0·60423	0·58201	0·56070	0·54027	0·50187	0·46651	0·40388	0·32690	0·23257
9	0·59190	0·56735	0·54393	0·52158	0·50025	0·46043	0·42410	0·36061	0·28426	0·19381
10	0·55839	0·53273	0·50835	0·48519	0·46319	0·42241	0·38554	0·32197	0·24718	0·16151
11	0·52679	0·50021	0·47509	0·45134	0·42888	0·38753	0·35049	0·28748	0·21494	0·13459
12	0·49697	0·46968	0·44401	0·41985	0·39711	0·35553	0·31863	0·25668	0·18691	0·11216
13	0·46884	0·44102	0·41496	0·39056	0·36770	0·32618	0·28966	0·22917	0·16253	0·09346
14	0·44230	0·41410	0·38782	0·36331	0·34046	0·29925	0·26333	0·20462	0·14133	0·07789
15	0·41727	0·38883	0·36245	0·33797	0·31524	0·27454	0·23939	0·18270	0·12289	0·06491
16	0·39365	0·36510	0·33873	0·31439	0·29189	0·25187	0·21763	0·16312	0·10686	0·05409
17	0·37136	0·34281	0·31657	0·29245	0·27027	0·23107	0·19784	0·14564	0·09293	0·04507
18	0·35034	0·32189	0·29586	0·27205	0·25025	0·21199	0·17986	0·13004	0·08081	0·03756
19	0·33051	0·30224	0·27651	0·25307	0·23171	0·19449	0·16351	0·11611	0·07027	0·03130
20	0·31180	0·28380	0·25842	0·23541	0·21455	0·17843	0·14864	0·10367	0·06110	0·02608
25	0·23300	0·20714	0·18425	0·16398	0·14602	0·11597	0·09230	0·05882	0·03038	0·01048
30	0·17411	0·15119	0·13137	0·11422	0·09938	0·07537	0·05731	0·03338	0·01510	0·00421
35	0·13011	0·11035	0·09366	0·07956	0·06763	0·04899	0·03558	0·01894	0·00751	0·00169
40	0·09722	0·08054	0·06678	0·05542	0·04603	0·03184	0·02209	0·01075	0·00373	0·00068
45	0·07265	0·05879	0·04761	0·03860	0·03133	0·02069	0·01372	0·00610	0·00186	0·00027
50	0·05429	0·04291	0·03395	0·02689	0·02132	0·01345	0·00852	0·00346	0·00092	0·00011
55	0·04057	0·03132	0·02420	0·01873	0·01451	0·00874	0·00529	0·00196	0·00044	0·00004
60	0·03031	0·02286	0·01726	0·01305	0·00988	0·00568	0·00328	0·00111	0·00023	0·00002

TABLE 3. PRESENT VALUE OF AN ANNUITY

The present value of £1 per annum for n years when discounted at interest rate r per annum $= \{(1 - (1 + r)^{-n})/r\}$. (See note C)
The amount per annum to redeem a loan of £1 at the end of n years and provide interest on the outstanding balance at r per annum can be determined from the reciprocals of values in this table. (See note D)

Interest % (=100r)

n (years)	1	1·5	2	2·5	3	3·5	4	4·5	5	5·5
1	0·9901	0·9852	0·9804	0·9756	0·9709	0·9662	0·9615	0·9569	0·9524	0·9479
2	1·9704	1·9559	1·9416	1·9274	1·9135	1·8997	1·8861	1·8727	1·8594	1·8463
3	2·9410	2·9122	2·8839	2·8560	2·8286	2·8016	2·7751	2·7490	2·7232	2·6979
4	3·9020	3·8544	3·8077	3·7620	3·7171	3·6731	3·6299	3·5875	3·5460	3·5052
5	4·8534	4·7826	4·7135	4·6458	4·5797	4·5151	4·4518	4·3900	4·3295	4·2703
6	5·7955	5·6972	5·6014	5·5081	5·4172	5·3286	5·2421	5·1579	5·0757	4·9955
7	6·7282	6·5982	6·4720	6·3494	6·2303	6·1145	6·0021	5·8927	5·7864	5·6830
8	7·6517	7·4859	7·3255	7·1701	7·0197	6·8740	6·7327	6·5959	6·4632	6·3346
9	8·5660	8·3605	8·1622	7·9709	7·7861	7·6077	7·4353	7·2688	7·1078	6·9522
10	9·4713	9·2222	8·9826	8·7521	8·5302	8·3166	8·1109	7·9127	7·7217	7·5376
11	10·3676	10·0711	9·7868	9·5142	9·2526	9·0015	8·7605	8·5289	8·3064	8·0925
12	11·2551	10·9075	10·5753	10·2578	9·9540	9·6633	9·3851	9·1186	8·8633	8·6185
13	12·1337	11·7315	11·3484	10·9832	10·6350	10·3027	9·9856	9·6829	9·3936	9·1171
14	13·0037	12·5434	12·1062	11·6909	11·2961	10·9205	10·5631	10·2228	9·8986	9·5896
15	13·8650	13·3432	12·8493	12·3814	11·9379	11·5174	11·1184	10·7395	10·3797	10·0376
16	14·7179	14·1313	13·5777	13·0550	12·5611	12·0941	11·6523	11·2340	10·8378	10·4622
17	15·5622	14·9076	14·2919	13·7122	13·1661	12·6513	12·1657	11·7072	11·2741	10·8646
18	16·3983	15·6725	14·9920	14·3534	13·7535	13·1897	12·6593	12·1600	11·6896	11·2461
19	17·2260	16·4262	15·6785	14·9789	14·3238	13·7098	13·1339	12·5933	12·0853	11·6077
20	18·0455	17·1686	16·3514	15·5892	14·8775	14·2124	13·5903	13·0079	12·4622	11·9504
25	22·0231	20·7196	19·5234	18·4244	17·4131	16·4815	15·6221	14·8282	14·0939	13·4139
30	25·8077	24·0158	22·3964	20·9303	19·6004	18·3920	17·2920	16·2889	15·3725	14·5337
35	29·4086	27·0756	24·9986	23·1452	21·4872	20·0007	18·6646	17·4610	16·3742	15·3906
40	32·8347	29·9158	27·3555	25·1028	23·1148	21·3551	19·7928	18·4016	17·1591	16·0461
45	36·0945	32·5523	29·4902	26·8330	24·5187	22·4954	20·7200	19·1563	17·7741	16·5477
50	39·1961	34·9997	31·4236	28·3623	25·7298	23·4556	21·4822	19·7620	18·2559	16·9315
55	42·1472	37·2715	33·1748	29·7140	26·7744	24·2641	22·1086	20·2480	18·6335	17·2252
60	44·9550	39·3803	34·7609	30·9087	27·6756	24·9447	22·6235	20·6380	18·9293	17·4500

Interest % (=100r)

n (years)	6	6·5	7	7·5	8	9	10	12	15	20
1	0·9434	0·9390	0·9346	0·9302	0·9259	0·9174	0·9091	0·8929	0·8696	0·8333
2	1·8334	1·8206	1·8080	1·7956	1·7833	1·7591	1·7355	1·6901	1·6257	1·5278
3	2·6730	2·6485	2·6243	2·6005	2·5771	2·5313	2·4869	2·4018	2·2832	2·1065
4	3·4651	3·4258	3·3872	3·3493	3·3121	3·2397	3·1699	3·0373	2·8550	2·5887
5	4·2124	4·1557	4·1002	4·0459	3·9927	3·8897	3·7908	3·6048	3·3522	2·9906
6	4·9173	4·8410	4·7665	4·6938	4·6229	4·4859	4·3553	4·1114	3·7845	3·3255
7	5·5824	5·4845	5·3893	5·2966	5·2064	5·0330	4·8684	4·5638	4·1604	3·6046
8	6·2098	6·0888	5·9713	5·8573	5·7466	5·5348	5·3349	4·9676	4·4873	3·8372
9	6·8017	6·6561	6·5152	6·3789	6·2469	5·9952	5·7590	5·3282	4·7716	4·0310
10	7·3601	7·1888	7·0236	6·8641	6·7101	6·4177	6·1446	5·6502	5·0188	4·1925
11	7·8869	7·6890	7·4987	7·3154	7·1390	6·8052	6·4951	5·9377	5·2337	4·3271
12	8·3838	8·1587	7·9427	7·7353	7·5361	7·1607	6·8137	6·1944	5·4206	4·4392
13	8·8527	8·5997	8·3577	8·1258	7·9038	7·4869	7·1034	6·4235	5·5831	4·5327
14	9·2950	9·0138	8·7455	8·4892	8·2442	7·7862	7·3667	6·6282	5·7245	4·6106
15	9·7122	9·4027	9·1079	8·8271	8·5595	8·0607	7·6061	6·8109	5·8474	4·6755
16	10·1059	9·7678	9·4466	9·1415	8·8514	8·3126	7·8237	6·9740	5·9542	4·7296
17	10·4773	10·1106	9·7632	9·4340	9·1216	8·5436	8·0216	7·1196	6·0472	4·7746
18	10·8276	10·4325	10·0591	9·7060	9·3719	8·7556	8·2014	7·2497	6·1280	4·8122
19	11·1581	10·7347	10·3356	9·9591	9·6036	8·9501	8·3649	7·3658	6·1982	4·8435
20	11·4699	11·0185	10·5940	10·1945	9·8181	9·1285	8·5136	7·4694	6·2593	4·8696
25	12·7834	12·1979	11·6536	11·1469	10·6748	9·8226	9·0770	7·8431	6·4641	4·9476
30	13·7648	13·0587	12·4090	11·8104	11·2578	10·2737	9·4269	8·0552	6·5660	4·9789
35	14·4982	13·6870	12·9477	12·2725	11·6546	10·5668	9·6442	8·1755	6·6166	4·9915
40	15·0463	14·1455	13·3317	12·5944	11·9246	10·7574	9·7791	8·2438	6·6418	4·9966
45	15·4558	14·4802	13·6055	12·8186	12·1084	10·8812	9·8628	8·2825	6·6543	4·9986
50	15·7619	14·7245	13·8007	12·9748	12·2335	10·9617	9·9148	8·3045	6·6605	4·9995
55	15·9905	14·9028	13·9400	13·0836	12·3186					
60	16·1614	15·0330	14·0392	13·1594	12·3766					

TABLE 4. SINKING FUND

The amount per annum for n years at interest rate r per annum needed to accumulate to £1 $= \{r/((1+r)^n - 1)\}$. (See note E)
The amount to which £1 per annum will increase in n years when accumulated at interest rate r per annum can be determined from the reciprocals of values in this table. (See note E)

Interest % $(=100r)$

n (years)	1	1·5	2	2·5	3	3·5	4	4·5	5	5·5
1	1·00000	1·00000	1·00000	1·00000	1·00000	1·00000	1·00000	1·00000	1·00000	1·00000
2	0·49751	0·49628	0·49505	0·49383	0·49261	0·49140	0·49020	0·48900	0·48780	0·48662
3	0·33002	0·32838	0·32675	0·32514	0·32353	0·32193	0·32035	0·31877	0·31721	0·31565
4	0·24628	0·24444	0·24262	0·24082	0·23903	0·23725	0·23549	0·23374	0·23201	0·23029
5	0·19604	0·19409	0·19216	0·19025	0·18835	0·18648	0·18463	0·18279	0·18097	0·17918
6	0·16255	0·16053	0·15853	0·15655	0·15460	0·15267	0·15076	0·14888	0·14702	0·14518
7	0·13863	0·13656	0·13451	0·13250	0·13051	0·12854	0·12661	0·12470	0·12282	0·12096
8	0·12069	0·11858	0·11651	0·11447	0·11246	0·11048	0·10853	0·10661	0·10472	0·10286
9	0·10674	0·10461	0·10252	0·10046	0·09843	0·09645	0·09449	0·09257	0·09069	0·08884
10	0·09558	0·09343	0·09133	0·08926	0·08723	0·08524	0·08329	0·08138	0·07950	0·07767
11	0·08645	0·08429	0·08218	0·08011	0·07808	0·07609	0·07415	0·07225	0·07039	0·06857
12	0·07885	0·07668	0·07456	0·07249	0·07046	0·06848	0·06655	0·06467	0·06283	0·06103
13	0·07241	0·07024	0·06812	0·06605	0·06403	0·06206	0·06014	0·05828	0·05646	0·05468
14	0·06690	0·06472	0·06260	0·06054	0·05853	0·05657	0·05467	0·05282	0·05102	0·04928
15	0·06212	0·05994	0·05783	0·05577	0·05377	0·05183	0·04994	0·04811	0·04634	0·04463
16	0·05794	0·05577	0·05365	0·05160	0·04961	0·04768	0·04582	0·04402	0·04227	0·04058
17	0·05426	0·05208	0·04997	0·04793	0·04595	0·04404	0·04220	0·04042	0·03870	0·03704
18	0·05098	0·04881	0·04670	0·04467	0·04271	0·04082	0·03899	0·03724	0·03555	0·03392
19	0·04805	0·04588	0·04378	0·04176	0·03981	0·03794	0·03614	0·03441	0·03275	0·03115
20	0·04542	0·04325	0·04116	0·03915	0·03722	0·03536	0·03358	0·03188	0·03024	0·02868
25	0·03541	0·03326	0·03122	0·02928	0·02743	0·02567	0·02401	0·02244	0·02095	0·01955
30	0·02875	0·02664	0·02465	0·02278	0·02102	0·01937	0·01783	0·01639	0·01505	0·01381
35	0·02400	0·02193	0·02000	0·01821	0·01654	0·01500	0·01358	0·01227	0·01107	0·00997
40	0·02046	0·01843	0·01656	0·01484	0·01326	0·01183	0·01052	0·00934	0·00828	0·00732
45	0·01771	0·01572	0·01391	0·01227	0·01079	0·00945	0·00826	0·00720	0·00626	0·00543
50	0·01551	0·01357	0·01182	0·01026	0·00887	0·00763	0·00655	0·00560	0·00478	0·00406
55	0·01373	0·01183	0·01014	0·00865	0·00735	0·00621	0·00523	0·00439	0·00367	0·00305
60	0·01224	0·01039	0·00877	0·00735	0·00613	0·00509	0·00420	0·00345	0·00283	0·00231

Interest % (=100r)

n (years)	6	6·5	7	7·5	8	9	10	12	15	20
1	1·00000	1·00000	1·00000	1·00000	1·00000	1·00000	1·00000	1·00000	1·00000	1·00000
2	0·48544	0·48426	0·48309	0·48193	0·48077	0·47847	0·47619	0·47170	0·46512	0·45455
3	0·31411	0·31258	0·31105	0·30954	0·30803	0·30505	0·30211	0·29635	0·28798	0·27473
4	0·22859	0·22690	0·22523	0·22357	0·22192	0·21867	0·21547	0·20923	0·20027	0·18629
5	0·17740	0·17563	0·17389	0·17216	0·17046	0·16709	0·16380	0·15741	0·14832	0·13438
6	0·14336	0·14157	0·13980	0·13804	0·13632	0·13292	0·12961	0·12323	0·11424	0·10071
7	0·11914	0·11733	0·11555	0·11380	0·11207	0·10869	0·10541	0·09912	0·09036	0·07742
8	0·10104	0·09924	0·09747	0·09573	0·09401	0·09067	0·08744	0·08130	0·07285	0·06061
9	0·08702	0·08524	0·08349	0·08177	0·08008	0·07680	0·07364	0·06768	0·05957	0·04808
10	0·07587	0·07410	0·07238	0·07069	0·06903	0·06582	0·06275	0·05698	0·04925	0·03852
11	0·06679	0·06506	0·06336	0·06170	0·06008	0·05695	0·05396	0·04842	0·04107	0·03110
12	0·05928	0·05757	0·05590	0·05428	0·05270	0·04965	0·04676	0·04144	0·03448	0·02526
13	0·05296	0·05128	0·04965	0·04806	0·04652	0·04357	0·04078	0·03568	0·02911	0·02062
14	0·04758	0·04594	0·04434	0·04280	0·04130	0·03843	0·03575	0·03087	0·02469	0·01689
15	0·04296	0·04135	0·03979	0·03829	0·03683	0·03406	0·03147	0·02682	0·02102	0·01388
16	0·03895	0·03738	0·03586	0·03439	0·03298	0·03030	0·02782	0·02339	0·01795	0·01144
17	0·03544	0·03391	0·03243	0·03100	0·02963	0·02705	0·02466	0·02046	0·01537	0·00944
18	0·03236	0·03085	0·02941	0·02803	0·02670	0·02421	0·02193	0·01794	0·01319	0·00781
19	0·02962	0·02816	0·02675	0·02541	0·02413	0·02173	0·01955	0·01576	0·01134	0·00646
20	0·02718	0·02576	0·02439	0·02309	0·02185	0·01955	0·01746	0·01388	0·00976	0·00536
25	0·01823	0·01698	0·01581	0·01471	0·01368	0·01181	0·01017	0·00750	0·00470	0·00212
30	0·01265	0·01158	0·01059	0·00967	0·00883	0·00734	0·00608	0·00414	0·00230	8·46 −4*
35	0·00897	0·00806	0·00723	0·00648	0·00580	0·00464	0·00369	0·00232	0·00113	3·39 −4*
40	0·00646	0·00569	0·00501	0·00440	0·00386	0·00296	0·00226	0·00130	5·62 −4*	1·36 −4*
45	0·00470	0·00406	0·00350	0·00301	0·00259	0·00190	0·00139	7·36 −4*	2·79 −4*	5·47 −5*
50	0·00344	0·00291	0·00246	0·00207	0·00174	0·00123	8·59 −4*	4·17 −4*	1·39 −4*	2·20 −5*
55	0·00253	0·00210	0·00174	0·00143	0·00118					
60	0·00187	0·00152	0·00123	0·00099	0·00080					

* The figures −4 or −5 indicate that the figures preceding them should be multiplied by 10^{-4} or 10^{-5}.

Bibliography

ALFRED A. M. and EVANS J. B. *Discounted cash flow.* Chapman and Hall, 1965.

ASHTON T. S. *The Industrial Revolution, 1760–1830.* Oxford University Press, 1958.

ASHWORTH W. *A short history of the international economy, 1850–1950.* Longmans, 1959.

BAUER P. T. and YAMEY B. S. *The economics of underdeveloped countries.* Nisbet and Cambridge University Press, 1957.

CARSBERG B. V. *Introduction to mathematical programming for accountants.* Allen and Unwin, 1969.

CARSBERG B. V. and EDEY H. C. (editors). *Modern financial management.* Penguin, 1969.

CEGB. *An appraisal of the technical and economic aspects of Dungeness B Power Station.* Central Electricity Generating Board, 1965.

CLARK C. *The conditions of economic progress.* Macmillan, 1957.

CLARK C. *The economics of irrigation.* Pergamon, 1967.

The financial and economic obligations of the nationalized industries. HMSO, 1961.

GRAAF J. DE V. *Theoretical welfare economics.* Cambridge University Press, 1957.

GRANT E. L. and IRESON W. G. *Principles of engineering economy.* Ronald Press, 1960.

HOGG V. W. Feasibility studies: an international lender's view. *Conference on Civil Engineering Problems Overseas 1966.* Instn civ. Engrs.

IBRD. *The World Bank—policies and operations.* International Bank for Reconstruction and Development, 1968.

LEWIS W. A. *Development planning.* Allen and Unwin, 1966.

LEWIS W. A. *Theory of economic growth.* Allen and Unwin, 1955.

LUCE R. D. and RAIFFA H. *Games and decisions.* Wiley, 1957.

MASSÉ P. *Optimal investment decisions.* Prentice-Hall, 1962.

MCKEAN R. N. *Efficiency in government through systems analysis: with emphasis on water resource development.* Wiley, 1958.

MERRETT A. J. and SYKES A. *The finance and analysis of capital projects.* Longmans, 1963.

MERRETT A. J. and SYKES A. *Capital budgeting and company finance.* Longmans, 1966.

181

MORONEY M. J.   *Facts and figures.*   Penguin, 1953.

NATIONAL ECONOMIC DEVELOPMENT COUNCIL.   *Investment appraisal.*   HMSO, 1969.

Nationalized industries: a review of economic and financial objectives.   HMSO, 1967.

PAISH F. W.   *Business finance.*   Pitman, 1968.

PREST A. R. and TURVEY R.   Cost–benefit analysis: a survey.   *Surveys of economic theory.*   Macmillan, 1967.

Proposed practices for economic analysis of river basin projects.   Report of the International Agency Committee on Water Resources, Washington, 1958.

REID P. A.   Appraisal of irrigation projects in underdeveloped countries.   *5th Congress International Commission on Irrigation and Drainage, Tokyo, 1963.*

SCHLAIFER R.   *Probability and statistics for business decisions.*   McGraw-Hill, 1959.

SOLOMON E.   *The theory of financial management.*   Columbia University Press, 1963.

TAK H. G. VAN DER.   *The economic choice between hydroelectric and thermal power development.*   World Bank, 1966.

TURVEY R.   *Optimal pricing and investment in electricity supply.*   Allen and Unwin, 1968.

VICKERS A.   The engineer in society.   *Proc. Instn mech. Engrs,* 1968–69, 183, Pt I, No. 5.

WEILLE J. DE.   *Quantification of road user savings.*   World Bank, 1966.

WEINGARTNER H. M.   *Mathematical programming and the analysis of capital budgeting problems.*   Prentice-Hall, 1964.